Growth Market Nanotechnology

Edited by
Norbert Malanowski, Thomas Heimer, Wolfgang Luther, and Matthias Werner

Related Titles

Junghans, C., Levy, A.

Intellectual Property Management

A Guide for Scientists, Engineers, Financiers, and Managers

188 pages with approx. 20 figures
2006
Hardcover
ISBN 3-527-31286-2

Fecht, H.-J., Werner, M. (eds.)

The Nano-Micro Interface

Bridging the Micro and Nano Worlds

351 pages with 102 figures and 27 tables
2004
Hardcover
ISBN 3-527-30978-0

Heimer, T., Werner, M.

Die Zukunft der Mikrosystemtechnik

Chancen, Risiken, Wachstumsmärkte

207 pages with 57 figures and 19 tables
2004
Hardcover
ISBN 3-527-31053-3

Bamfield, P.

Research and Development Management in the Chemical and Pharmaceutical Industry

280 pages with 57 figures and 19 tables
2003
Hardcover
ISBN 3-527-30667-6

Growth Market Nanotechnology

An Analysis of Technology and Innovation

Edited by
Norbert Malanowski, Thomas Heimer,
Wolfgang Luther, and Matthias Werner

WILEY-VCH

WILEY-VCH Verlag GmbH & Co. KGaA

The Editors

Dr. Norbert Malanowski
VDI TZ GmbH
Future Technologies Division
Graf-Recke-Str. 84
40239 Düsseldorf
Germany

Prof. Dr. Thomas Heimer
Hochschule für Bankwirtschaft
Sonnenmannstr. 9–11
60314 Frankfurt
Germany

Dr. Wolfgang Luther
VDI TZ GmbH
Future Technologies Division
Graf-Recke-Str. 84
40239 Düsseldorf
Germany

Dr. Matthias Werner
NMTC
Rüsternallee 29
14050 Berlin
Germany

Translation
Christine Ahner
translate.economy
Frhr.-v.-Eichendorff-Str. 8/1
88239 Wangen im Allgäu
Germany
translate.economy@web.de

Cover
Extruder for a "Nanotechnology Desktop Factory" as proposed by John Burch and Eric Drexler, www.lizardfire.com

Library of Congress Card No.: applied for
British Library Cataloging-in-Publication Data:
A catalogue record for this book is available from the British Library.

Bibliographic information published by Die Deutsche Bibliothek
Die Deutsche Bibliothek lists this publication in the Deutsche Nationalbibliografie; detailed bibliographic data is available in the Internet at http://dnb.ddb.de.

Typesetting TypoDesign Hecker GmbH, Leimen
Printing Strauss GmbH, Mörlenbach
Binding Litges & Dopf GmbH, Heppenheim

Printed in the Federal Republic of Germany
Printed on acid-free paper

ISBN-13: 978-3-527-31457-7
ISBN-10: 3-527-31457-1

Table of Contents

Growth Market Nanotechnology. Edited by N. Malanowski, T. Heimer, W. Luther, and M. Werner

ISBN: 3-527-31457-1

Foreword

The Wave of the Future

Nanoscience and nanotechnology have opened an era of integration of fundamental research and engineering from the atomic and molecular levels, enhancing innovation for economic manufacturing of products, human health and cognitive technologies in the long term. For this reason, nanotechnology R&D investments by governments worldwide have increased tenfold since 1997 to over $4 billion in 2005 and the private sector has allocated another $4 billion in the same year. Now, all Fortune 500 companies in advanced materials, electronics and pharmaceuticals have R&D activities in this field.

While R&D activities have spread rapidly in the last five years (they are now in over 60 countries), and nanotechnology has been recognized as a key R&D domain, the economic potential and societal benefits of nanotechnology basically remain in the exploratory phase. Rather, they are seen as promising, and the national investment policies do not generally recognize nanotechnology to be as important as information technology and biotechnology. This may be explained by the relatively recent developments in nanoscale knowledge and the limited economic understanding of its implications. There is an apparent gap between the accelerated accumulation of scientific data, and ways to apply the results safely and economically. The promise for future economic benefits is a key driver for any emerging technology, but it is generally difficult to document it. This publication addresses an important need for evaluating technological and economic implications of nanotechnology in the international context. The data show that Germany holds a strong position in Europe and the world alongside the United States and Japan. Information from government and private sources are presented with a wealth of data on economic trends, patents and publication, and application areas.

Despite its importance, there is no internationally recognized definition for nanotechnology. Any definition of nanotechnology should include three elements:

Growth Market Nanotechnology. Edited by N. Malanowski, T. Heimer, W. Luther, and M. Werner

ISBN: 3-527-31457-1

- the size range of the structure (which is generally defined as the intermediate length scale between a single atom or molecule, and about 100 molecular diameters or about 100 nm; this condition alone is not sufficient because all natural and manmade systems have a structure at the nanoscale);
- the ability to work at that scale (without the ability to measure and transform at the nanoscale we do not have new understanding and a new technology; such ability has been reached only partially so far, and significant progress has been made in the last five years);
- exploitation of properties and functions specific for the nanoscale, compared with the macro or micro scales (this is the motivation for researching at the nanoscale).

According to the National Science Foundation and National Nanotechnology Initiative, nanotechnology is the ability to understand, control, and manipulate matter at the level of individual atoms and molecules, as well as at the "supramolecular" level involving clusters of molecules. Its goal is to create materials, devices, and systems with essentially new properties and functions because of their small structure. The definition implies using the same principles and tools to establish a unifying platform for science and engineering at the nanoscale, and using the atomic and molecular interactions to develop efficient manufacturing methods.

The authors of this volume provide useful data on existing nanotechnology applications, and attempt to predict the next steps. These developments must be placed in the long-term context. Much of what has already made it into the market place is in the form of "first generation" passive nanostructures, but many small and large companies have "second" and embryonic "third generation" products in the pipeline, while concepts for the "fourth generation" products are still only in research (M.C. Roco, *J. Nanoparticle Res.*, 2005). The *First Generation* (after 2000) of passive nanostructures is illustrated by nanostructured coatings, nanoparticles, nanowires, and bulk nanostructured materials. The *Second Generation* (after 2005) of active nanostructures is illustrated by transistors, amplifiers, targeted drugs and chemicals, sensors, actuators, and adaptive structures. The *Third Generation* (predicted after 2010) includes three-dimensional nanosystems and systems of nanosystems using various synthetic and assembling techniques such as bio-assembling; nanoscale robotics; networking at the nanoscale and multiscale architectures. The *Fourth Generation* (after 2015–2020) includes heterogeneous molecular nanosystems, where each molecule in the nanosystem has a specific structure and plays a different role. Molecules will be used as devices, and fundamentally new functions will emerge from their engineered structures and architectures. This book covers especially "first" and the beginning of the "second generation" where more data are available, and raises issues related to the future generations of nanotechnology products.

January 2006 *Mihail C. Roco*
National Science Foundation
and National Nanotechnology Initiative
Arlington, Virginia, USA

Contributors

Dr Gerd Bachmann studied physics at the University of Kaiserslautern and worked for several years in the Department of Physics there, where his work was mainly concerned with surface analysis. In 1992 he completed this scientific work with a thesis on "Ion bombardment induced effects in Ta/Si multilayer systems". Since 1991 he has been a consultant at VDI (Association of Engineers), Düsseldorf in the Future Technologies Division of the VDI Technology Centre (VDI TZ). VDI TZ works mainly on behalf of the German research ministry (BMBF), combining funding tasks with forecasting and assessment topics. Gerd Bachmann is engaged in the analysis and evaluation of future technologies, mainly in the areas of nanotechnology, nanobiotechnology, X-ray technologies and lithography. In the field of nanotechnology he works mainly on strategies for promoting innovation processes via funding support.

Prof Dr Thomas Heimer is the Dean of HfB / Business School of Finance and Management at Frankfurt-am-Main, Germany. Before starting his career at HfB he held the position as head of the Research Department at VDI/VDE-IT in Teltow, where he was primarily involved in research in innovation, financing and emergence of new companies. Thomas Heimer worked at the Johann Wolfgang Goethe University and was a visiting researcher at Colombia University in New York, where he acquired teaching experience in Economics, Innovation Theory and Human Resource Management.

Dr Andreas Hoffknecht joined the Future Technologies Division of VDI TZ as a technology consultant in 2000. He studied physics at the University of Giessen, Germany. From 1996 to 2000 he held the position of a scientific assistant at the private University of Witten/Herdecke, Germany. He is a specialist in the field of technology forecasting and is the co-ordinator of the project "Technology Forecasting for the German Federal Ministry of Education and Research". His par-

Growth Market Nanotechnology. Edited by N. Malanowski, T. Heimer, W. Luther, and M. Werner

ISBN: 3-527-31457-1

ticular interests include applied physics, electronics, information technology, self-assembling and nanotechnology.

Dr Dirk Holtmannspötter has been a Senior Consultant and Project Manager at the Future Technologies Division of the VDI TZ since 1999. He holds a Master of Science Degree (1991), a Diploma (1993) and a PhD (2002), all in Physics. His current work focuses on photonics, optoelectronics and nanotechnology. His methodological expertise lies in technology forecasting, bibliometrics and patent analysis. He has completed a number of projects for industry, the German Federal Ministry of Research and Education and for the EU in these fields.

Dr Wolfgang Luther has been working as a technology consultant at the Future Technologies Division of VDI-TZ in Düsseldorf since 1999. His current main field of activity is the coordination of innovation-promoting measures for nanotechnology within the funding programme of the German Federal Ministry of Research and Education. He holds a degree in Chemistry (1995), a degree in Economics (1999) and a PhD in Analytical Chemistry (1998). He worked as scientific assistant at the Federal Institute for Materials Research and Testing (1995–1998) and as a consultant for the Institute Logos GmbH (1999). His specific fields of competence are technology forecasting and the assessment of emerging technologies, in particular nanotechnology. He has completed a number of projects for the EU, the German Ministry of Education and Research and for various industrial clients in this field.

Dr Norbert Malanowski has been working as a Senior Visiting Scientist with the European Commission, Information and Communication Technologies Unit at the Institute for Prospective Technological Studies (IPTS), since September 2005. His current work focuses on Converging Applications for Active Ageing. He has a Diploma in Social Sciences (1988, University of Duisburg and University of Toronto) and a PhD in Political Science/Political Economy (1996, University of Hagen). He worked as a consultant for different private and public organisations (1990–96) and has been a lecturer at the Open University in Hagen, Germany since 1991. He was a Senior Researcher at the University of Bochum, Germany 1996/97, and at the Center for Technology Assessment in Stuttgart from 1998/99. Since 1999 he has been a Senior Consultant and Project Manager at VDI TZ in Düsseldorf. His specific fields of competence are converging applications, nanotechnology, ICT and labor market policy and ageing societies. Norbert Malanowski has completed a number of projects for the EU, several German Federal Ministries and industrial clients in these fields.

Stephan Mietke received his Dipl.-Ing. in Engineering and Business from the Technical University of Berlin, Germany and his Master's degree in Management from the Stevens Institute of Technology, New Jersey, USA. He started his professional career with Deutsche Bank AG in 1997, where he had worked in different corporate banking functions in Berlin and Frankfurt am Main before he joint Deutsche Bank's Microtechnology Innovation Team, where he advised and assessed start-up and established technology companies from a technical and financial point of view. From 2004 to 2005 he worked as business consultant on nano- and microtechnologies, focussing on strategic consulting of companies and public authorities in these emerging fields. In 2005, Stephan left the micro- and nanotechnology arena to join the Association of German Banks, where he deals with retail banking and banking technology issues. He has (co-)authored several scientific papers and book chapters and has given invited talks.

Dr Hermann Sanders has been working as a consultant to GOBABEB Research and Training Centre, Namibia since 2005. From 1995 to 2004 he was a consultant to the Government of Burgenland, Austria and to private companies. He also worked in research projects on nanotechnology at HfB / Business School of Finance and Management at Frankfurt am Main, Germany. Hermann Sanders studied economics and social sciences at the universities of Münster, Hamburg, Basel and Regensburg.

Dr Volker Wagner works as a technology consultant with VDI TZ, Düsseldorf, Germany. After his PhD at the Max Planck Institute for Chemistry in Mainz, Germany, he worked as project scientist at the University of Leeds, UK, managing a European Research Project on chemical kinetics from 1999 to 2001. Since 2001 he has been working at VDI TZ on a variety of projects within the German Nanotechnology campaign of the German Research Ministry. He has analyzed the current research activities in the field of nanobiotechnology and nanomedicine for German governmental bodies and for the European Commission. Apart from his consulting work in the nanobiotechnology arena, Volker Wagner joins a technology-watch task force that advises the German Research Ministry on emerging technologies in the fields of chemistry, physics and nanotechnology.

Dr Matthias Werner is Managing Director of NMTC, a consulting company focusing on strategic consulting of companies and public authorities in the emerging fields of nano- and microtechnology. In 1998 he became Director of the Microtechnology Innovation Team of Deutsche Bank AG. The focus of the Microtechnology Innovation Team was on technical consulting, market analysis and the evaluation of high-tech companies in the field of nano- and microtechnology in Germany. His previous occupations included consulting at VDI/VDE-IT GmbH where he managed federal research programmes. He received his Dipl.-Ing. and Dr.-Ing. degrees in electrical engineering from the Technical University of Berlin and has published more than 90 peer-reviewed scientific papers, book chapters, books and reports and has given numerous invited lectures at international conferences. Matthias Werner is the only foreign expert so far to have become a member of the UK National Strategic Advisory Group (NSAG).

Dr Dr Axel Zweck has been head of the Future Technologies Division of VDI TZ since 1993. His current main research are questions of the development of emerging technologies, particularly of nanotechnology, and the use of weak signal detection of technologies, risks and trends. He holds a degree in Chemistry (1987), a Master in Social Sciences (1988), a PhD in Biochemistry (1989) and a PhD in Social Sciences (1992). He worked as Scientist at the Department of Biochemistry at the University of Düsseldorf (1987–1989) and as consultant at VDI TZ (1989–1993). Since 2002 he has held a teaching position at the University of Düsseldorf. His areas of expertise are Research and Technology Management, Technology Assessment, Technology Forecasting, Foresight and Evaluation. Axel Zweck was and is project manager on different projects on emerging technologies, sustainability and forecasting for the EU, various German Federal Ministries and industrial clients.

Overall Synopsis

Norbert Malanowski

- Nanotechnology is one of the key technologies of the 21st century. These days, considerable turnover is achieved with products that are only feasible by the existence of nanotechnology. With the economic breakthrough of nanotechnology, this turnover should increase significantly in the future.
- The few publications available that deal with the direct and indirect economic potential of nanotechnology contain significant weaknesses. However, they offer a useful basis for the present study, which is intended to be a realistic assessment of the market volume and market relevance, both for Germany itself and in the international context.
- An internationally standardized definition of nanotechnology has not yet been established. Even the issue of differentiating between nanotechnology and microtechnology or existing chemical processes, or the association of nanotechnology with different procedures and methods, has to be explained by examples. An absolutely precise and unambiguous definition for the classification of technological processes and products as nanotechnology does not exist.
- Within the context of this study, nanotechnology describes the creation, analysis and application of structures, molecular materials, inner interfaces and surfaces with at least one critical dimension or with manufacturing tolerances (typically) below 100 nanometers. The decisive factor is that new functionalities and properties resulting from the nanoscalability of system components are used for the improvement of existing products or the development of new products and application options. These new effects and possibilities are predominantly based on the ratio of surface-to-volume atoms and on the quantum-mechanical behavior of the elements in the material.
- Currently, there are about 450 "nanotechnology companies" in Germany. In this study, the term "nanotechnology company" refers to companies that participate in projects of the Federal Ministry of Education and Research (BMBF) in the field of nanotechnology and to

Growth Market Nanotechnology. Edited by N. Malanowski, T. Heimer, W. Luther, and M. Werner

ISBN: 3-527-31457-1

those companies classified as such by the authors of this study because of their activities in the field of nanotechnology.

- Evaluation of the patent situation firmly establishes that the extremely dynamic development of nanotechnology is reflected in patent applications. During the past five years, the annual number of patent applications in nanotechnology has doubled about every two years.
- Analysis of figures for patent applications across different countries shows that Germany is in an excellent position, both in nanotechnology as a whole, and in the quantitatively most important section of chemistry. Germany's position appears to be particularly good regarding the especially valuable patents. In none of the Lead Markets (chemistry, car manufacture, and optics) that were surveyed in the patent analysis is Germany lagging seriously behind the USA or Japan.
- When discussing patent strategies in the field of nanotechnology, industry experts agree that, in principle, nanotechnology does not require completely different patent strategies to other fields of technology. A characteristic, however, is the fact that in certain cases the period of development from a nanotechnological basic effect to its application can be so long that patent protection expires shortly after the maturity stage of the product has been reached.
- Industry experts point out that in small and middle-sized enterprises (SME) the tendency to patent is far more modest than in large-scale enterprises. If they are uncertain, SME prefer not to disclose technological trade secrets at all – even if they were patentable. For financial reasons, SME are more likely to register utility patents, for example. According to the experts canvassed, a reasonable patent strategy for SME could be to have a technology patented together with the main users.
- We must emphasize that a search with such a wide topical scope as that presented here, is only appropriate for a general assessment of the patent situation. The assessment of individual companies and specific technologies requires more detailed patent searches.
- It is not possible to evaluate the "world market for nanotechnology" on the basis of the figures given in publicly available studies, since market figures are only available for a part of the nanotechnological products, thus lists are incomplete. In addition, market forecasts partly refer to different time horizons; nanotechnological products are named twice in two or more sections (e.g. application of basic nanoproducts/components in end products of different branches of industry); and products from different stages of the value-added chain are taken into account (basic products, intermediate products, end products etc.). A summary of the market potential of the most important nanotechnological applications in different

branches is given in tabular form in Chapters 3 (Table 3.9) and 6 (Table 6.8).

- Comparison of different market forecasts for the world market for nanotechnology from different sources shows that the leverage effect caused by nanotechnology affects a world market of currently 100 billion EUR. Here, it should be recognized that market volumes of various value-added depths have been added. On average, the market forecasts predict an exponential increase over the next ten years. Examination of the individual market shares attributable to Germany is not simple. It is universally accepted that nanotechnology will play a positive role in the future development of German enterprises.
- For the medium term, the electronics market will still be dominated by CMOS-technology. Until 2006, the share of nanoelectronics (i.e. feature widths <100 nm) will amount to 10% of the total CMOS-market with a world market volume of approx. 20 billion USD. Magnetoelectronics has already conquered significant market shares in the field of hard disc drives in the form of GMR-read heads and will offer the potential to introduce MRAM-chips in the DRAM-chip market on a medium-term basis.
- In the field of chemistry, long-established nanostructured materials, such as carbon black, silicic acid or polymer dispersions, realized turnovers in the order of billions on the world market, albeit with only a slight market growth. Dynamic market growth is expected for more recent nanomaterials, such as carbon nanotubes, polymer nanocomposites, aerogels, organic semiconductors and inorganic nanoparticles, provided that "show-stoppers" of a technological (e.g. problems with the upgrading of manufacturing processes) or socio-ecological nature (e.g. toxicity of nanomaterials) do not impede progress.
- Currently, German nanotechnology companies and car manufacturers regard the market relevance of nanotechnology in car manufacture as relatively low, inter alia, because of the long lead times of technological developments, coupled with innovation cycles of different model series. However, nanotechnology has already found its way into the production of some car components (e.g. scratch-proof paint, nano-covered injection pumps, LED tail lights etc.). In the long run, nanotechnological know-how will represent a decisive competitive advantage in car manufacture, covering all relevant criteria from ecology (e.g. energy-efficient drives, light-weight manufacture, reduction of pollutants and the protection of natural resources) to safety (passive and active safety) and to comfort (product design, infotainment etc.).
- Market potential in the optical industry mainly arises in the field of ultraprecise optics for semiconductor manufacturing (optical li-

thography or photolithography), in the field of optoelectronic light sources (laser diodes and LED) and in the display sector (OLED and FED), where turnovers in the order of billions of USD are expected until the year 2006.

- In the long run, medicine and life sciences are considered one of the major markets for nanotechnology. In the medium term , the sales potential of nanotechnological products is considered to be low compared with those of the chemical industry, optics and electronics, for example. The main part of the nanotechnological market potential in the field of life sciences is based on biomedical quick tests (DNA, protein chips), where the influence of nanotechnology is felt first and foremost in the sector of detection systems.
- The results of the company survey show that the starting point for nanotechnology in enterprises lay mainly in the period from 1996 to 2000. Over this period of time, observation of the nanotechnology scene, some R&D, and the exploitation of nanotechnology for products, all experienced a considerable increase. Within this period, discussions took place about market relevance (1996) and publically-funded competence centers for the nanotechnology sector were established (1998).
- The enterprises polled within the framework of this study disagree with the statement that nanotechnology only represents a new experimental field, just as they reject, in a more moderate way, the statement that through nanotechnology, technological competence would be extended. This result clearly refutes the popular opinion that nanotechnology is simply a "hype".
- Furthermore, the results of the company survey showed that the chemistry sector (including materials) clearly heads the list of nanotechnology companies and applications in Germany (judging by the number of enterprises, the frequency of already existing nanotechnological products and their sales potential up to the year 2006), followed by life sciences (medical technology/health) and information and communication technology (ICT).
- The most serious obstacles to innovation in Germany are high investment costs, lack of loan capital and insufficient financial support. The ranking order of the obstacles mentioned indicates that the development of new products or procedures in the nanotechnology sector requires considerable investment that cannot be financed by equity capital alone. The development of markets with the help of nanotechnology also entails remarkable investments that industry alone is not able to raise easily. Limited market knowledge and networks of cooperation that are as yet undeveloped, are obstacles to the speed of innovation and the diffusion of new application fields, especially in branches which are still not penetrated that much by nanotechnology. Cooperation between public finance

and nanotechnology companies also represents a significant challenge for the future, a problem not yet solved especially in Germany.

- Important differences between the obstacles to innovation for SME and large-scale enterprises can be identified in three sections:
  1. There is a clear difference with regard to the sources of finance. SME consider their access to the capital market to be more difficult than is the case of large-scale enterprises. Accordingly, the financing of their activities constitutes an important obstacle to innovation for 38.6% of the SME. In contrast to this, access to the capital market for large-scale enterprises represents a serious obstacle to innovation for only 7.7% of the sample.
  2. The difference is similar in relation to access to market information. With 21.3%, here again the number of SME considering this to be a significant obstacle to innovation is much higher than the number of the large-scale enterprises, with only 3.7%.
  3. Finally, the lack of competent regional cooperative partners is another obstacle to innovation evaluated differently by SME and large-scale enterprises in the sample: 22.2% of the SME consider this obstacle to be important, whereas only 7.4% of the large-scale enterprises state this as an important obstacle to innovation.
- The withdrawal of venture-capital investment from start-up enterprises has extremely negative effects on the setting-up of businesses in Germany in the nanotechnology environment at present. The classical bank financing of business start-ups has become more and more difficult over recent years, because the German banking landscape itself is going through a crisis.
- According to conservative assessments of the employment growth in nanotechnology, an increase of at least 10 000 to 15 000 jobs is expected until 2006. Although it is not possible to give an accurate prediction of the number of jobs in the field of nanotechnology, we already know that about 20 000 to 32 000 (conservative assessment) and 114 000 (optimistic assessment) jobs in about 450 enterprises in Germany are directly or indirectly dependent on nanotechnology
- By 2015 it is expected that nanotechnology will affect almost every sector of industry. As expected, the fields influenced most by nanotechnology are, seen from an international point of view, the chemical industry, life sciences and electronics.
- In conclusion we can say that at the moment, Germany has an excellent basis for the economic realization of its nanotechnological activities. Its excellence in research, however, is not completely reflected by economic realization. In this area the USA and Japan are superior to Germany. We also have to take into account that investment and government funding in the field of nanotechnology have increased considerably all over the world. This can be put down to

the very high market volumes predicted. Therefore, even stronger international competition with regard to nanotechnology can be expected in the future.

1
Introduction

Norbert Malanowski and Axel Zweck

1.1
Initial Situation and Goal of the Study

Key technology

Nanotechnology is one of the key technologies of the 21st century. These days, considerable turnover is achieved with products that are only feasible by the existence of nanotechnology. With the economic breakthrough of nanotechnology, this turnover should increase enormously in future.

Basic hypothesis

Examples for current and future fields of application of nanotechnology are, inter alia, pharmaceuticals, electronics, the field of new materials, car manufacture, mechanical engineering and environmental technology. However, a generally recognized extensive database of the economic potential has not yet been developed, nor is one publicly available. Therefore, at present the economic potential of nanotechnology cannot be assessed realistically by means of inquiries based on the purely quantitative methods of empirical economic research. This is the basic hypothesis of the present study.

Existing studies fragmentary

Lead Markets

Major studies published so far (e.g. Evolution Capital 2001; DG Bank and GZ Bank, 2001; Beckmann and Lenz, 2002; TAB, 2003) are still too fragmentary to be able to portray precisely the economic importance of nanotechnology for all lines of industry concerned, particularly as the definition remains vague. Moreover, these major studies do not usually consider sufficiently the so-called Lead Markets[1] of the individual countries. The fields of electronics, information and communication technology (ICT) and biotechnology, for example,

1) In literature, the term "Lead Market" is used differently at times. In this study, we prefer a definition proposed in a study by the Federal Ministry of Education and Research in 2002, which states: "Lead Markets are regional markets (usually countries) using a certain innovation design earlier than other countries and incorporating specific features (Lead Market Factors) that increase the probability of the same innovation design being adopted broadly in other countries." (Federal Ministry of Education and Research, 2002, p. 108).

Growth Market Nanotechnology. Edited by N. Malanowski, T. Heimer, W. Luther, and M. Werner

ISBN: 3-527-31457-1

represent significant Lead Markets in the USA, whereas in Germany the Lead Markets are mainly the chemical industry, car manufacture, optics, life sciences (in this study medical technology/health) and electronics. These Lead Market sectors are characterised by their particularly intensive relationship to science, from which they draw their technological strength (see, inter alia, BMBF, 2002; Beise, 2002; Managermagazin, 9/2002).

Although the few publications available that deal with the direct and indirect economic potential of nanotechnology contain significant weaknesses, they offer a useful basis for the present study, which is intended to be a realistic assessment of the market volume and market relevance, both for Germany itself and in the international context.

1.2 Methodological Procedure

Method mix

For such a study on the economic potential of a still very young, new technology, a method mix is just the right thing. According to Alemann (1995), qualitative (e.g. expert interviews and literature analyses) and quantitative methods (e.g. standardized surveys) should be regarded as complementary, rather than competitive, methods of gaining insights. For qualitative methods, open procedures are preferred that impose as few restrictions as possible on the persons questioned when formulating their subjective reality constructions. In the context of quantitative methods, standardized measuring instruments are usually applied to quantify the variables measured and analyze them using statistical methods. In the meantime, the conviction gaining more and more acceptance in the empirical economic and social research is that "there is no ideal methodological solution, but a combination of methods has to be applied, especially of qualitative and quantitative methods that take into account the subject area, the problem concerned and the financial and time resources available" (Wollmann, 2001, S. 382).

Proven its worth in practice

This procedure of combined methods, or mixed methods, has already proven its worth for the analysis of data material for the evaluation of microsystems engineering (another quite new technology) and can help to circumvent the deficits of each of the qualitative and quantitative methods in a study on the economic potential of nanotechnology (Heimer and Werner, 2004).

Work phase 1

After the determination of a (broad) definition of nanotechnology, at first intensive enquiries about market reports and patents concerning nanotechnology were carried out on the internet and in databases. In addition to that, approx. 15 exploratory expert interviews were conducted using an interview guide (see Appendix 1). This helped above all in gathering background information (work phase 1).

After this preparatory work, an enquiry into German companies active in the field of nanotechnology followed using a standardized questionnaire (see Appendix 2). Before its application, the questionnaire underwent a pre-test. The business survey was analyzed using the statistical program SPSS, because this software package offers sound evaluation possibilities and provides excellent data management (work phase 2).

Work phase 2

All the processed results of literature analysis, patent analysis, expert interviews and business survey were brought into expert workshops in the form of propositions as a constructive basis for confrontation (Delphi method[2]). Experts from banks, science, competence centers for nanotechnology, producers, suppliers, system developers and venture capital enterprises participated in these workshops. These experts not only served as critical commentators of the outcome of the literature analysis, expert interviews, patent analysis and business survey, but also contributed additional expert knowledge about the economic potential of nanotechnology from their own points of view. To be able to show the importance of nanotechnology for the German Lead Markets, special attention was paid to such lines of business that impact on the technological competitiveness of the German industry (chemistry, car manufacture, optics, medicine and life sciences[3], and electronics[4]). The data gained from this concentration was processed again by the project team and brought into the particular branch-specific expert network for validation before its final documentation (work phase 3).

Work phase 3

Final pooling and assessment of all determined results followed. The results have been assessed, inter alia, in the context of results from other available commercial sources (Markus-Database and cur-

Work phase 4

2) In the Delphi method, experts in the respective field to be examined are polled, as a rule in one or more runs. They are usually presented with an explicitly structured catalogue containing questions and propositions, on the basis of which they are expected to make their assessments of future developments and trends.

3) The BMBF had initiated a study on "Nanotechnology and Health" which was published in autumn 2004. This study also comprises results regarding the market potential of nanotechnology in the field of life sciences. In order to avoid the same work being carried out twice, it was decided not to hold a workshop in this subject.

4) Thanks to the International Technology Roadmap for Semiconductors (ITRS), electronics belong to the group of technologies whose future development is described as being very good. As the semiconductor industry is very investment-intensive, market forecasts are extremely important for entrepreneurial decisions. This demand is filled by a number of (commercial) market research institutes. Currently, the forecasts reach approx. to the year 2008. The question of what will be the development of the nanotechnological share compared to microelectronics, and which consequences are to be expected for which branches of industry has not yet been examined systematically enough, however, it could be elaborated by means of a secondary analysis of selected commercial market studies. Therefore, a special workshop was not necessary.

rent market study of the Deutsche Bank; Fecht et al., 2003). This was important to validate the results and to develop specific perspectives for the location Germany. Therefore a strength–weakness analysis of the international competitors in this sector, their current and future orientation in the field of research and application was made and compared with Germany's position. Based on the compiled information and with the help of structured methods (SWOT analysis and white spot analysis) (work phase 4), an analysis of the status in Germany and the existing opportunities and deficits was carried out.

1.3 Structure of the Report

nanotechnology: general notes

The present study is structured according to this multistage and combined methodological procedure. After the general notes on nanotechnology (inter alia, on the definition, new effects as well as on players and international activities) in Chapter 2, the next chapter describes above all the current and future applications and market prospects in products and product groups. Products in this context are interpreted, among other things, as products from the sectors of nanomaterials, nanoelectronics and nanooptics. A preliminary assessment of the market prospects was made, based on a secondary analysis of the available German and international literature (Chapter 3).

Secondary analysis

Patent analysis

On the basis of a patent analysis, statements are made about the priorities concerning the content and the temporal development of patent applications in nanotechnology worldwide (Chapter 4). Through the examination of information about inventors and patent applicants available in the relevant databases (e.g. WP-INDEX, EUROPATFUL, USPATFUL), knowledge is gained about the role of German enterprises and scientists in this field[5].

Company survey

Within the framework of the company survey, the results of which are shown in Chapter 5, it is determined, inter alia, at which points of the value-added chain entrepreneurs (suppliers and system developers) in Germany go into the sector of nanotechnology. This provides clues for the assessment of the economic prospects for the exploitation of nanotechnology. Parallel to this and within the context of the survey, prerequisites/competences are identified which are of great importance for the development, application and diffusion of nanotechnology. Finally, obstacles to nanotechnological innovation are shown. With this, three priorities of the company survey are determined: (market) potentials, prerequisites to the application and innovation/diffusion obstacles.

Three priorities of the survey

5) This study has been carried out with the special cooperation of the European Patent Office, contact: Manfred Scheu.

Regarding the market potential, the survey records how companies assess the effects of the application of nanotechnology, i.e. in view of international competitiveness (exports), growth prospects and possible direct employment effects.

The topic field of "prerequisite/competences" portrays especially the demands on the basic resources of enterprises (know-how, qualification of the staff etc.). Possible obstacles to the economic application of nanotechnology may result from deficiencies in basic resources, problems regarding cooperation with scientific institutions and may extend to problems concerning acceptance by potential customers and lack of infrastructural preconditions.

The delimitation of the nanotechnological market and the identification of the enterprises active in this sector serve as a base for a first empirical assessment of the employment effects triggered off by this new technology and its products in Germany. The current primary employment effect can be derived from the total number of employees in small and middle-sized enterprises and large-scale enterprises.

The survey was directed at selected companies already active in the field of nanotechnology. The selection of these companies was based on three sources: at first, companies within the sphere of competence centers for nanotechnology were included. This circle was extended to companies taking or having taken part in development programs for nanotechnology. Finally, these two sources were complemented by young enterprises on the basis of data provided by Venture Capital Enterprises, the Deutsche Bank and similar institutions. From these sources, a stock of approx. 450 enterprises to be polled was obtained.

Approx. 450 companies working in nanotechnology

Since nanotechnology is still in an early phase of commercial exploitation, i.e. its prospects are developing only gradually, from an economic research point of view such a result-oriented parent population is clearly superior to a purely random sample of companies. However, the specific selection of the companies has to be taken into consideration when interpreting these data (see Chapter 5 for more details).

The processing of the acquired data (from work phases 1 and 2) with regard to a clear definition of relevant markets and market volumes, an identification of affected lines of business, a classification of Germany's foreign trade position and an assessment of the employment effects as well as the positioning of Germany in the international context, served as input for branch-specific workshops in the form of a catalogue with questions and propositions (in the sense of a constructive confrontation basis). To be able to demonstrate the importance of nanotechnology for German Lead Markets, the products identified were correlated with the respective markets. Data about Germany's international position in the nanotechnological sector were acquired by means of self-assessments of the enterprises within the

Workshops specific to lines of business

context of the company survey[6]. The companies use current information to predict market potentials expected in the near future and the position of their international competitors with their respective products. The results of this work process are dealt with in Chapter 6.

Market Assessment The total results of the work phases 1–4 are combined, discussed and subjected to a market assessment in Chapter 7. For this reason, the results are assessed in context with results of other available commercial sources. Here, among other things, Germany's competitive position in the international environment is evaluated qualitatively.

Conclusions In Chapter 8 the results are summarized and options for action for players from business, science and politics are derived from the outcome of this study.

Acknowledgements

This study was drawn up within the context of the project "Innovation and Technical Analysis on Nanotechnology" carried out by the Future Technologies Division of the VDI Technology Center GmbH on behalf of and with support of the Federal Ministry of Education and Research (BMBF), Department LS 21 (Support Code 16 I 1503). The type of problem was predetermined by the BMBF. The Federal Ministry of Education and Research had no influence on the outcome of the study; the contractor takes the sole responsibility. The authors are responsible for the content. The opinions expressed do not necessarily correspond to the opinion of the Federal Ministry for Education and Research.

We want to express our thanks to a large number of experts for their valuable contributions and suggestions. Naming all of them would go beyond the scope of this page. Special thanks go to the participants of the project workshops for their commitment.

6) Foreign trade data for innovative products of the international product nomenclature are mostly calculated with the help of the Innovation Global Sourcing Management Tools (GSMT) on the basis of OECD or EUROSTAT (e.g. ZEW). This is possible for lines of industry covered by these statistics. Currently, this tool is only partly applicable to nanotechnology as a technology in a nascent state, since turnovers, research & development costs etc. are recorded only insufficiently in government statistics like EUROSTAT.

2 Notes on Nanotechnology

Gerd Bachmann

2.1 Definition of a Complex Term

Nanotechnology is gaining increasing public attention worldwide and is described as one of the most important future technologies. However, it is less a basic technology in the classical sense with a clear definition, than a new interdisciplinary approach applicable to different lines of business for further progress in electronics, optics and biotechnology or for new materials. On the one hand, in nanotechnology, engineering with elementary units of active and inactive nature, i.e. atoms and molecules, is applied as if working with a Lego kit. On the other hand, even structures measuring only one thousandth of the diameter of one hair can be created by means of reduction. This problem is comparable to the challenge of writing the whole road network of Germany on a fingernail true to scale – and faultlessly, of course. In the product sector, however, an evolutionary rather than revolutionary development is taking place at the moment. Computers are getting faster and faster, mobile phones are more and more versatile, lenses in DigiCams are smaller and smaller and car paint, for example, becomes harder and harder. Here, nanotechnology provides growing insights. It is likely that in the future, market sectors will change in a revolutionary way; for example in the pharmaceutical and medical sector or, in the not too distant future, in lighting engineering.

No clear definition

In the work of the Federal Ministry of Education and Research (BMBF) in the field of nanotechnology, the following phraseology is used in official publications (2004):

Use of practicable phraseology

"Nanotechnology describes the creation, examination and application of structures, molecular materials, inner interfaces and surfaces with at least one critical dimension or with manufacturing tolerances (typically) below 100 nanometers. The decisive factor is that new functionalities and features for the improvement of existing products or the development of new products and application options result from

Growth Market Nanotechnology. Edited by N. Malanowski, T. Heimer, W. Luther, and M. Werner

ISBN: 3-527-31457-1

the nanoscalability of the system components alone. These new effects and possibilities are predominantly based on the ratio of surface to volume atoms and on the quantum-mechanical behavior of the elements of the material."

An internationally standardized definition of nanotechnology has not yet been established. Even the issue of differentiating between nanotechnology and microtechnology or existing chemical processes, or the association of nanotechnology with different procedures and methods, has to be explained by examples. An absolutely precise and unambiguous definition for the classification of technological processes and products as nanotechnology does not exist.

Nanotechnology versus nanotechnology

A further problem emanates from the "precision" of the possible definition. More application-oriented experts define nanotechnology as a field with dimensions smaller than microbiology, with a broad grey area belonging to one of the two fields, whereas the "hardliners" are advocates of "molecular nanotechnology"; atom per atom and molecule per molecule, they examine the application of individual elements (atoms and molecules) to create systems.

Thus, difficulties arise when it comes to the objective description of what nanotechnology really is. This becomes clear in the presentation of the industrial products portrayed in this study, because nano-relevance is not always clearly attributable to a product. The following discussion makes clear why the working definition from the BMBF is both pragmatic and useful.

Size alone is insufficient

The two basic criteria for the definition of nanotechnology are often used differently. On the one hand, purely geometrical criteria can be applied considering only the size of the objects. One nanometer is a billionth of a meter (10^{-9} m), approx. 50 000 times smaller than the diameter of one human hair, or only ten times larger than a hydrogen atom. On this scale, the laws of classical physics no longer apply, but new properties and functionalities arise because very small particles have a large ratio of surface atoms to volume atoms and because of the quantum-mechanical behavior of the elements of the material (see Fig. 2.1).

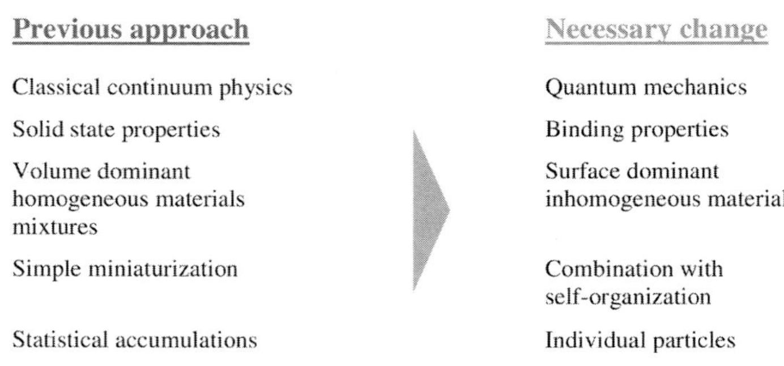

Figure 2.1 Changed approach of selected features in transition to the nanoscale.

This indicates that the geometrical criterion alone is not sufficient to limit the scope of nanotechnology. The second criterion, necessary for this reason, is based on a phenomenological point of view. With nanoscalability, new effects, functionalities and new qualities of the properties arise (e.g. antireflex properties, transparency, scratch-resistance, colorfulness etc.). These new physical properties, however, may depend not only on one generally determinable particle size, but also on the material class, for instance. Moreover, nanotechnology provides new possibilities of manufacturing and control for individual objects on the nanoscale. This includes self-organizing systems that establish a new structure from nanoscale elements (e.g. molecules).

These two aspects of nanotechnology require a coupled consideration of structural dimensions and functions as described in the wording of the BMBF definition. The reason why a precise definition of what constitutes nanotechnology turns out to be difficult (and wouldn't make sense) is the existence of numerous borderline cases. Most of the research programs worldwide use examples and comparisons of quantities to give an idea of what should be classed as nanotechnology.

Coupled structural dimensions and functions

2.2
Bottom-up and Top-down Strategies

Changes of properties caused by nanoscaling are to a great extent based on the new approach of using dimension, form and composition to gain new physical, chemical and biological active principles. Dependent on these integrative tendencies, today's nanotechnology has developed mainly from three directions that meet on the nanolevel (see Fig. 2.2):

Three lines of development

- In recent decades, physical–technical procedures were the driving forces for the generation of increasingly complex circuits, and with this, of smaller structures (top-down efforts) in microelectronics. On the shop shelves we find processors of increasingly higher power, as well as memory modules and hard discs with increasing capacities.
- Insights gained from complex chemistry and supramolecular chemistry led to the specific establishment of high-molecular and functional chemical compounds with an enormous application potential in catalysis, membrane technology, sensor or layer technology (bottom-up efforts).
- Recently, the comprehension of biological processes on a cellular and molecular level has increased decisively. This includes a multitude of processes of technologically unequalled functionality and

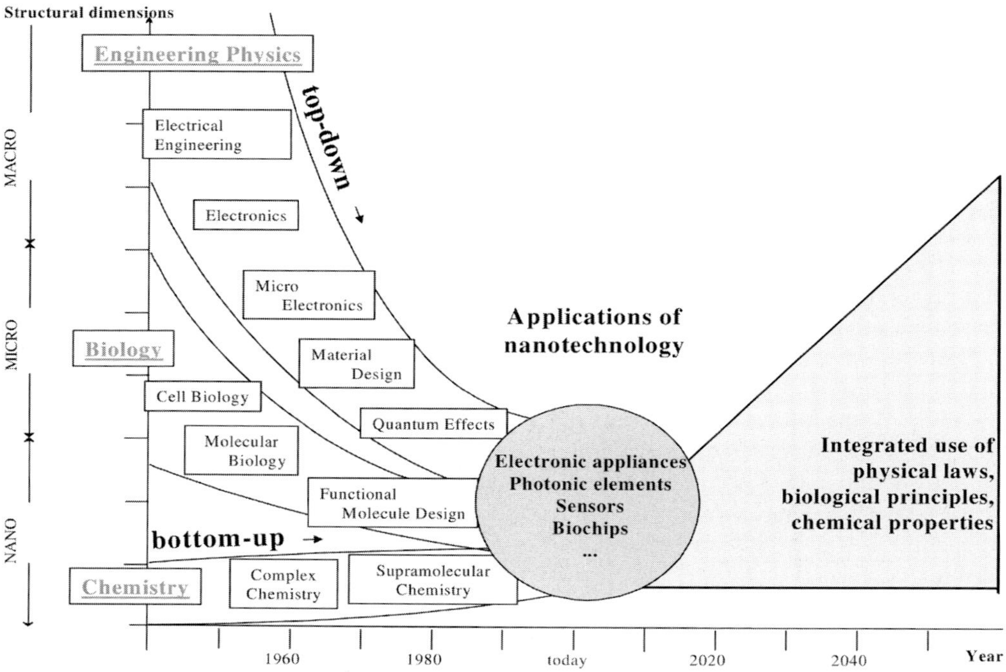

Figure 2.2 General development trends and relation to nanotechnology (source: VDI TZ).

complexity in the most confined of spaces, such as self-organization of molecular compounds or photosynthesis. In future it will be necessary to enhance the transfer of basic biological principles to technical systems. At the same time, biotechnology provides a more and more complex toolbox of procedures for the design of functional molecules that may bring the future application of biological–technical hybrid systems within our reach, e.g. for implants, artificial muscles or the replacement of organs.

Methods of one discipline may also be completed adequately by procedures and specialized knowledge from other subject areas. Physical processes are usually used to examine nanoscale objects or to carry out specific structuring. The generation of nanoscale particles, however, is first and foremost a domain of chemistry. Biological nanoobjects, such as proteins, enzymes or viruses, develop by means of self-organization according to the construction plans of nature, with the majority of the basic processes, such as photosynthesis, happening on the nanoscale or molecular level.

2.3
New Effects through Nanoscalability

An atom or molecule does not yet display physical properties familiar to us, such as electric conductivity, magnetism, color, mechanical strength or a certain melting point. However, materials with the size of a dust particle do possess all those physical properties, which do not differ from those of a heavyweight steel object. Thus, nanotechnology occurs in a transient area between individual atoms or molecules on the one hand and larger solids on the other hand. In this twilight zone, phenomena occur that are not observed on macroscopic objects (Fig. 2.3). *Nanoscale effects*

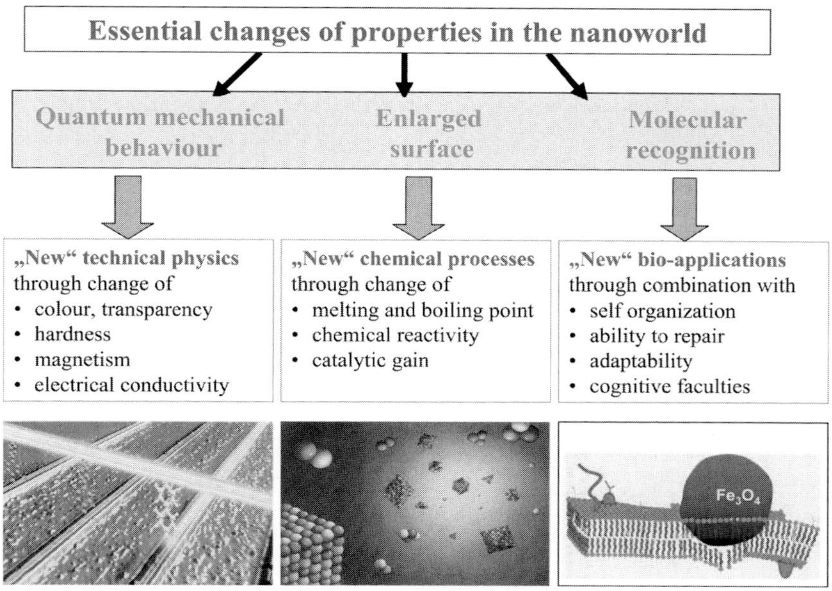

Figure 2.3 Essential changes of properties in the nanoworld.

Some examples of new functionalities:

- The increasing complexity of information technology requires new electronic and optoelectronic features that are only possible by using smaller components.
- The smallest particles offer new fields of application for lacquers and paints, such as various color effects through controlled change of particle size, or transparent but still functional coatings, such as dirt-repellent seals or UV-protection.

- Minimal admixtures of nanomaterials can significantly change the properties of a solid, so that plastic films become wear-proof and ceramics almost unbreakable.
- The chemical reactivity and working life of catalytic converters can clearly be increased through adequate structure compositions on a surface.

2.4 Thematic and Structural Interdisciplinarity

To promote nanotechnology it is first necessary to expand nanotechnological know-how through interdisciplinary approaches in research and development by concentrating existing scientific resources. Then the application-oriented conversion of these concepts into marketable products is an inevitable task in a highly developed national economy. Nanotechnology is a typical cross-disciplinary technology and the development of new applications is at the core of many combined projects in the specialized programs of the BMBF (2002). Such programs aim to extract identifiable application prospects from the basic results of nanotechnology research and to support enterprises that generate added-value in the realization of these results. Therefore, in nanotechnology it is important to unite the players qualified for certain applications on an interdisciplinary basis and to take the necessary steps along the added-value chain in order to improve the market opportunities of those involved in the innovation process of international scientific and economic competition.

BMBF supports interdisciplinary approach

In this process, "stamina" and the readiness for constant further development are required, as it is often necessary to go beyond the limits of current disciplines and to enter into new untested partnerships.

The essential content of such collaborations are given below.

2.4.1 Nanomaterials, Ultrathin Coatings and Porous Structures

Nanoparticles display an enormous application spectrum because of their increased reactivity, and they allow targeted functional combination with various chemical derivatives. Special functions are achieved, for example, by dispersion and stabilization of these particles, for instance in the form of liquid formulations of low viscosity, high-filled ceramic-slip, transparent multifunction coatings, pigment dispersions, e-inks and ferrofluids. Special surface-modified magnetic nanoparticles are under research to mark and combat tumor cells. The coating of nanoparticles will contribute to the improved handling of sensitive nanomaterials or to the protection from chemical reactions. Among other things, this concerns composites of nanoparticles and plastics as well as nanocrystals, e.g. with optical properties.

Versatile nanoparticles

Nanostructural surface treatment is gaining in importance. This refers to any kind of coating that contributes significantly to the improvement of properties, such as scratch resistance or water and dirt resistance. Another important and technically relevant field of research is the improvement of anti-reflective properties and UV-protection for electro-chromic and photo-active coatings as well as the development of innovative abrasion-resistant coatings.

Another priority is the development of switchable, addressable or structurable thin coatings for technical applications by means of modifying magnetic properties, transparency or adjustable and controllable hydrophobia/hydrophilia. Photovoltaic coatings and the targeted development of switchable and renewable nanocoatings and nanostructures for microelectronic components, polymer electronics and smart polymers, displays, light and heat management as well as dampers, actuators and sensors are all of interest.

Functional coatings

Mesoporous or foam-like materials can considerably increase the power of fuel cells, batteries or storage media through the application of nanostructured mesoporous materials. Further fields of application will develop in the insulation of buildings, sound insulation, metal protection in motor vehicles and in the manufacturing of artificial paper.

A future focus of support and development will be the field of functional coatings. Product predictions in this sector of ultrathin, optically functional coatings include switchable mirrors, highly efficient thin film solar cells on the basis of quantum dots, and photo-addressable polymer films. New research fields can be developed in the sectors of light-activated plastic magnets and photosensitive magnetic switches on molecular basis.

2.4.2 Nanobiotechnology

The general objective of nanobiotechnology is the creation of interfaces between biological and technical systems on the biologically relevant scale of individual molecules and molecular compounds. Thus both the design of technical systems for the analysis and control of biological systems and the exploitation of biological systems or principles in technology are addressed.

Learning from natural processes

Current problems of nanobiotechnology occur particularly in the control of biological–technical interfaces, called "interface engineering" or interface design. The controlled handling of cells and cell compounds requires suitable nanostructured and functionalized surfaces and membranes. Apart from the field of tissue engineering, the pharmacological sector will profit most. It is foreseeable that interface design will become an important element of innovative techniques for the in-vivo validation of drug targets. Here, the goal is the provision of improved methods for faster and more specific testing or validation of active substances. The active functionalization of cells and tissue particles is important for future biohybrid systems. Fields of application are mainly neuro-active implants, research and/or treatment of neurodegenerative diseases as well as neurotechnology. The interconnection of electronic and biological systems is a key factor here and success in this field is an essential precondition to open the door to neurotechnology.

For the exploitation of biological materials and processes in engineering, the technical use of self-organization phenomena will be of special importance in future, inter alia as a possible alternative to conventional lithographic methods, as in the development and application of nanodimensional machine technologies. This includes the broad field of cell-free models of motion (e.g. protein motors) that may be used for nanoscale manipulation, controlled motion of objects or delivery of substances. The fields of application are mainly found in the biotechnological, biomedical and chemical sector.

For all fields mentioned, efficient and high-resolution analytical methods are indispensable. Here, the limits of optical, mechanical, chemical and biosensory methods have not yet been reached, nor have the limits of combinations of these methods.

2.4.3 Nanooptics

Ultrasmooth and structured optics

The term "nanooptics" is mainly used for the ultraprecise processing of optical components, in which the reproducible and economical production of optical components with accuracies below one nanometer is in the foreground. Such precision optics are mainly used in lithography, which is urgently required for the manufacture of electronic components with ever-diminishing structural sizes. In lithography, the quality of optics has top priority and is the limiting factor in what is currently technically feasible; this fact results in high manufacturing costs and thus in a high price. Furthermore, with increasingly smaller wavelengths it is necessary to use mirror optics instead of transmitting lens systems. Here, extremely complex coating systems are applied to achieve the necessary functionality of optics.

Far less drastic are the demands in the consumer field, as for example in the case of aspheres for data projectors, cameras, lenses, scanners etc. In this area the need is for efficient manufacturing and measurement of optics at moderate precision.

A relation to nanotechnology is also found in completely new optical concepts, such as "photonic crystals", in which a so-called band gap for light is feasible by means of suitable micro- and nanostructuring that allows the light to be led to a very confined space and to be manipulated, which is the key to microoptical integration.

Photonic crystals with a certain band gap require sufficiently regular and trouble-free structuring of a material in the range of the wavelengths of the conducted radiation as well as an efficient fiber launch. Based on III–V-semiconductors, they can be integrated with active components and so they open up the real vision of optoelectronics. Integrated optical switching matrices, add/drop-multiplexer and cross-connects are conceivable that may contain additional new functionalities such as modulators, monitor diodes or wavelength stabilizers. Microlaser with fiber launch, integrated polarization beam splitter and adjustable dispersion compensators are early examples. Advantages in performance and price make components based on photonic crystals extremely attractive for the broadband networks of the future.

Photonic crystals

2.4.4
Nano-optoelectronics and Spin Systems

LED

Nanotechnological aspects are also inherent in new types of semiconductor light sources (laser and light-emitting diodes [LED]). These optoelectronic components generate light in extremely thin semiconductor layers measuring only nanometers, and in individual cases even in quantum dot structures. This is a rare example of a bottom-up approach, rather than the miniaturization of an already known technology, that led to the introduction of new products on the market and to a huge economic success. However, even this field still requires intensive research for the development of new wavelength ranges, the improvement of light output, efficiency and life cycle.

With progressing miniaturization, the components of communications engineering have reached dimensions in which new physical effects emerge within the semiconductor structures that can be explained using quantum theory. The challenge is to develop new principles of the circuit design that use the quantum effects in the nano range which have to be taken into account there.

Use of quantum effects

It has become clear that the III–V-quantum structures are perfectly suitable for the examination of interaction phenomena and new kinds of collective electronic states in solids. Efforts are made to use these effects to manufacture transistors, LEDs and lasers, which are key elements of communications technology.

In the focal field of development of "electron correlations and dissipation processes in III-V-semiconductors", several research groups are working closely together to master the properties of materials and the manufacturing technology of such structures and to examine quantum effects.

Apart from components in which the electron charge is deployed for the transmission of information, quantum systems are examined in which the electron spin could be used and specifically manipulated. Therefore, the examination of spin-dependent phenomena and states and the realization of spin-components are also targets of the funded projects. The strong development of this field of basic research can also be deduced from the fact that the Nobel prizes for physics have been awarded repeatedly for the development of new components of information technology. Recipients have been Prof. v. Klitzing (D, Quanten-Hall-Effect), Prof. Stoermer (D/USA) and not least Prof. Kroemer (USA/D) who was given an award for the development of transistors with high electron mobility (HEMT).

2.4.5 Nanoelectronics

Nanofabrication

Within the framework of the production processes of nanoelectronics that are suitable for production, the manufacture of adequate structuring masks is one of the main objectives. Mask technology is an ultraprecision technology. Its contribution to nanoscale structural transfer in chip manufacturing is essential. The large-scale project of the BMBF on mask technology, running from 2003 until 2007, represents a central innovation for nanoelectronics. Mask technology will be researched for structures from 90 to 35 nanometers. In addition, alternative structuring techniques will be examined (switchable masks, maskless techniques).

The main trends of nanolithography are covered by the furthering of projects of 193 nm and EUVL. However, since EUVL does not necessarily cover the demand of chip manufacturers with great diversity and small number of pieces per type of chip (as it is mostly the case with ASICs, in power electronics and communication electronics), a parallel study on alternative structuring techniques is being carried out. Here, non-optical lithographic methods, such as electron and ion-beam lithography or new methods of replication are being analyzed with regard to their potential for the manufacture of future nanotechnological products. Moreover, for the more distant future, self-ordering processes are being discussed as possible structuring methods for the nanoscale.

The SOI-Technique (Silicon on Insulator) is a nanoscale modification of silicon on wafer basis. It contributes significantly to the control of the speed-reducing influences of the wafer and its insulation barriers.

For non-volatile memories four possible technological approaches regarding nanoelectronics are discussed (Flash-memory, MRAM, FRAM and phase change RAM). Further possible topics are new materials for gate dielectrics (so-called high-K materials), non-optical nanolithography, nanopackaging, assisted self-organization for nanoelectronics ("self-ordering"), 3D-structuring, programmable logic, and new concepts for the production engineering of nanoelectronics.

Future electronics

Within the context of the BMBF-project funding, topics such as spintronics, carbon nanotubes and molecular electronics are being supported. These projects are still of an exploratory nature. The objective is to examine suitable approaches for future industrial realization in Germany. Since December 2001, a variety of further emerging research devices for the period from 2012 has been described in the ITRS-Roadmap. Topics such as quantum-cellular machines, phase change memory and other matters are discussed there. For the most part, these fields are still in a state of basic research. Therefore, companies only observe them, but hardly ever research them. A new aspect of "emerging research devices" is that they often require know-how beyond the current expertise in microelectronics. For this reason, the development of infrastructural measures is discussed that links innovation players in the established field of microelectronics with those working on new topics. This allows a faster and more profound identification of the approaches worthy of investment within the framework of the BMBF-project funding at a later point in time.

2.4.6 Nanoanalysis

The eyes of the nanoworld

The main emphasis is currently on analytical processes in the application fields of biotechnology and semiconductor technology. They document the increasing orientation towards application even of the basic research related funding in this field. Following this trend, the topic field of "process relevance and suitability" will come to the fore in future. The technical focus could be on chemical-sensitive nanoanalysis, non-destructive analysis of hidden interfaces and highly-resolved analysis of large surfaces. In the medium term and with increasing industrial activity in nanotechnology, nanoanalysis and the field of standardization will be integrated in the fields of application.

2.4.7 Industrial Production

The potentials resulting from nanotechnology confront industry with the problem of putting R&D results immediately into practice so as to maintain and consolidate competitiveness on an international level.

From concept to product This transfer of results from a laboratory level into industrial practice represents an immediate obstacle that may be overcome by the cooperation of research institutes and industrial enterprises. Here, it is important to overcome the current limits of production technology in order to manufacture new and more efficient products in a process-reliable and economical way. Apart from technical challenges, problems regarding the organization of working processes and new demands on in-plant training and personnel development also have to be solved.

The methods of precision manufacturing currently applied have already reached accuracies of a few micrometers. Numerous procedures of microstructuring permit structural dimensions of this order of magnitude. The aim is to master new limits in the nanometer range (some hundred nm) using new procedures of both precision manufacturing and microstructural engineering. The same goes for assembly processes on which equally growing demands have to be placed. Here basic research work is necessary that requires the interplay of classical mechanical engineering and new methods of microsystems technology to overcome the barriers from "micro" to "nano".

New processes of surface coatings enable functional layers that are based on ultrathin layers with typical layer thicknesses of less than hundred nanometers. Such layers are mainly applied in the optical industry, for instance, for the manufacture of optical filters and lenses with defined spectral properties or for the application of functional coatings that protect large glass sheets from soiling. Here, the challenge is to apply layers, which are producible and reproducible on small surfaces, onto large surfaces in which tolerances of a few atomic layers have to be maintained.

In the production of nanomaterials, a compromise is still made between high quality, such as narrow particle size distribution of the material, and the high output rate of nanomaterials. The more limited the particle size distribution, the more effective are the typical features such as optical properties, magnetism or chemical reaction. The processing of these nanomaterials presents industry with a great challenge regarding the production of new kinds of materials like transparent ceramics with particular properties.

2.5 Central Players in Germany

Increasing comprehension of quantum effects, interface and surface properties and self-organizing principles, led in the past ten years, to the laying down of foundations for new analytical and production methods that triggered off a real boom in interest in nanotechnology

Infrastructure and resources

and network activities along the added-value chain all over the world. The early combination of results of basic research with application options and the expected market potential arising from it, spurred the interest enormously. The players in the nanotechnology scene in Germany were among the first in the world to address application options at an early stage on the basis of a sound and extensive basic research. About 450 enterprises in Germany have already recognized these chances of innovation and dedicate themselves more and more to this field of technology as product developers, suppliers or investors. They do not regard nanotechnological R&D work as a passing fashion, but dedicate themselves on a long term basis to the key elements for future new developments in lines of industry with high employment potential, mainly in the field of electronics, car manufacture, chemistry, optical manufacturing and life sciences or power generation and the building industry.

In Germany, the success of nanotechnology is based on a broad range of players from industry, science and politics. It would go beyond the scope of this study to describe this in detail, so the institutions mentioned in the following sections are representative of numerous players.

2.5.1 Networks

2.5.1.1 BMBF-funded Competence Centers (CCN)

BMBF competence centers for nanotechnology

In 1998, the BMBF established six competence centers with a development fund of about 2 million EUR per year. Since autumn 2003, nine competence centers continued or started their work as topical networks with regional clusters throughout Germany in the most important fields of nanotechnology:

- Ultrathin functional coatings (Dresden)
- Nanomaterials: Functionality through chemistry (Saarbrücken)
- Ultraprecise surface treatment (Brunswick)
- Nanobioanalytics (Münster)
- HanseNanoTec (Hamburg)
- Nanoanalytics (Munich)
- Nanostructures in Optoelectronics NanOp (Berlin)
- NanoBioTech (Kaiserslautern)
- NanoMat (Karlsruhe; established and financed by the Research Center Karlsruhe)

The aim of the infrastructural activity of the competence centers is to enable an optimum contact between potential users and nanotechnology researchers. At the same time, the nanotechnological expert knowledge of members can be concentrated efficiently and transferred into industrial development. Other tasks of the competence centers are in particular activities for job training and personnel development, cooperation in matters regarding standardization, advice and support for people willing to start up a business, and public relations. The individual competence centers are structured along thematic added-value chains in their respective field. About 440 players from universities, research centers, small and middle-sized enterprises (SME), financial service providers, consultants and associations, are currently organized in the whole network. The exchange of information organized by the competence centers mainly helps small enterprises to get information about current developments and to be able to assess it properly. In the following three years, the main focus will be on job training and personnel development as well as on the support of start-up businesses. The BMBF funding will be supplemented with the same amount by regional financing from the German Länder.

2.5.1.2 Other Networks

Apart from the competence centers financed directly by the BMBF, a number of networks with different goals and consequently even different constitutions have been established.

Other players in R&D

In contrast to the usually (virtually) designed networks throughout Germany, some universities and research centers concentrated their nanotechnological activities in the field of basic research by means of local – partly even internal – networks. Examples for this are the Center for Nanoscience, CeNS (Munich), the Center for Interdisciplinary Nanostructure Science and Technology, CINSAT (Kassel), the Center of Nanoelectronic Systems for Information Technology, CNI (Jülich) and the Center for Functional Nanostructures, CFN (Karlsruhe). NanoBioNet in the Saarland has a particularly regional character.

A special role is played by the establishment of incubators generated by universities and aiming at the support of hive-offs in the university environment. For this reason, the CeNTech GmbH from Münster (Center for Nanotechnology) has established its own start-up center.

2.5.2 Institutional Research Facilities

2.5.2.1 Wissensgemeinschaft G. W. Leibniz (WGL)

WGL

In the institutes of the WGL, both basic research-related and industry-oriented works in nanotechnology are carried out. The main emphasis is put on nanomaterials research; prominent players are the Institut für Neue Materialien (Institute for New Materials, INM, Saarbrücken); for Festkörper- und Werkstoffforschung (Solids and Materials Research, IFW, Dresden); and for polymer research (IPF, Dresden). Surface treatment is studied e.g. at the Institut für Oberflächenmodifizierung, (Institute for Surface Modification, IOM, Leipzig) and at the Forschungszentrum Rossendorf (Research Center Rossendorf, FZR). Basic work on solid-state electronics is carried out at the Paul-Drude-Institut (PDI, Berlin).

2.5.2.2 Helmholtz Gemeinschaft deutscher Forschungszentren (HGF) (Association of German Research Centers)

In the HGF the main emphasis is on work concerning material-relevant matters and on nanoelectronics. The activities of the two research centers in Karlsruhe (FZK) and Jülich (FZJ) are outstanding, and even the research center in Geesthacht (GKSS) and the Hahn-Meitner Institute in Berlin (HMI) carry out R&D on nanomaterials and coating systems.

HGF

2.5.2.3 Max-Planck-Gesellschaft (MPG)

Essential and basic insight into new approaches to nanotechnological research is provided by the work of the MPG-Institutes. The Stuttgart Institute for Solid-State Research and Metal Research and the MPI for Microstructure Physics in Halle have been working for many years in the fields of nanomaterials, supramolecular systems, characterization processes and new functionalities. R&D results recognized worldwide also come from activities of the Institutes for Polymer Research (Mainz), for Colloid and Interface Research (Golm), for Biochemistry (Munich-Martinsried), for Carbon Research (Mühlheim) and from the Fritz-Haber-Institut (Berlin).

MPG

2.5.2.4 Fraunhofer-Gesellschaft (FhG)

There is now an industrial demand in nearly all sections of nanotechnology, so the numerous Fraunhofer-Institutes are working in cooperation with industry on nanotechnology projects with specific application goals. The long-standing BMBF-funding supports the focal activities that are mainly found in the fields of coatings and surfaces; here the Institutes for Material and Beam Technology (IWS, Dresden), for Silicate Research (ISC, Würzburg), for Optics and Precision Engineering (IOF, Jena) and for Interface Research (IGB, Stuttgart) are very active. Nanomaterials research has priority, for example, in the Institutes for Applied Material Research (IFAM, Bremen), for Applied Solid-state Physics (IAF, Freiburg) and for Chemical Technology (ICT, Pfinztal). The transition from microtechnology to nanotechnology is studied by the Institutes for Silicon Technology (ISIT, Itzehoe) and for Production Technology (IPT, Aachen); nanobiotechnology is in the focus of interest for the Institute for Biomedical Technology (IBMT, St. Ingbert). The Institute for Solar Energies (ISE, Freiburg) examines the contribution of nanotechnology to the generation of energy.

FhG

2.5.3
Universities and Other Research Facilities

Universities

Nearly all German universities with technical and scientific course contents carry out nanotechnology-related R&D activities. As a result, interdisciplinary understanding of the connection between these sections is developing. Some universities have already established nanotechnology courses closely linked with the current research topics. The university locations of Karlsruhe, Aachen, Bielefeld, Munich, Hamburg, Saarbrücken, Kaiserslautern, Berlin, Kassel, Würzburg, Freiburg and Marburg are examples of the numerous activities covering almost the whole subject area. In addition, even the colleges of applied sciences are beginning to dedicate themselves more and more to this subject area (Cebulla, Malanowski and Zweck, 2005).

Apart from the institutes mentioned so far, further institutions with an emphasis on the field of nanotechnology exist in the strongly diversified R&D system in Germany, such as the AMICA in Aachen, NMI in Reutlingen, IMS-Chips in Stuttgart, FBI Berlin, Bessy II Berlin, PTB Braunschweig, CAESAR Bonn, IPHT Jena.

2.5.4
Industrial Research and Development

Industry

Currently, about 450 industrial enterprises are players in the field of nanotechnology in Germany (www.nano-map.de). In many large-scale enterprises, like Infineon, DaimlerChrysler, Schott, Carl Zeiss, Siemens, Osram, BASF, Bayer, Metallgesellschaft or Henkel, investigation of nanotechnology problems takes place in the R&D department. For instance, nearly all large chemical companies are dealing with the manufacture of nanoscale materials. Research may be organized in different ways: while Henkel, in cooperation with the Technical University of Darmstadt, spun off the SusTech company for the external development and marketing of new nanotechnological applications and materials and established it at a university, for Degussa the "Project House Nano", of the wholly-owned subsidiary Creavis, was responsible for in-house exploration of nanotechnological processes and products up to maturity of application and was supported by universities. These developments are currently transferred into business segments. A third model might be total outsourcing of the exploitation of results and patents. The company Sunyx, for example, has spun off from Bayer AG this way, or Mildendo from Jenoptik. Infineon AG is following another model, for which a corporate research department (Infineon-CPR Corporate Research), clearly related to nanotechnology, is given the task of realizing nanotechnological concepts. This company is specifically examining sub-50-nm-CMOS-transistors for future nanoelectronics and carbon nanotubes (CNT) as possible connections between different chip levels (chip interconnects).

Large-scale enterprises are more interested in system solutions with high sales prospects, whereas smaller and medium-sized companies (SME) are mainly committed to the fields of production analysis and instrument engineering. The company Nanogate Technologies GmbH (Saarbrücken), for example, is a SME that offers its nanomaterials for different application purposes (among other things for easy-to-clean-coatings, nonstick products, graffiti protection etc.). Also numerous start-up enterprises (hive-offs from universities and institutions) such as Nano-X, ItN-Nanovation, Capsulution etc. are important nanotechnology-players in Germany. Apart from purely material manufacturers, a lot of companies are working in nanostructuring (e.g. Aixtron, NaWoTec, Team Nanotech, Nanoplus) or nanoanalysis (among others Omicron Nanotechnologies, IoNTOF, NanoAnalytics, FRT, JPK, SIS, NanoTools).

2.6 Comparison of German Activities with the International Situation

2.6.1 Project Funding by the Public Sector

In Germany the support of nanotechnology by the public sector is mainly the responsibility of the Federal Ministry of Education and Research (BMBF) and the Federal Ministry of Economics and Technology (BMWi). Besides this, institutional research on nanotechnology outside the universities is in the hands of the German Research Society (DFG) and the four major public research groups: the WGL, HGF, FhG and MPG.

2.6.1.1 Federal Ministry of Education and Research (BMBF)

The BMBF has been funding research in the field of nanotechnology since the end of the 1980s within the context of the programs "Materials Research" and "Physical Technologies". At first, the manufacturing of nanopowders, the generation of lateral structures on silicon and the development of methods for nanoanalysis were the focal topics. Later on, nanotechnology-related research work was financed in other programs, too, such as "Laser Research" or "Optoelectronics". Today numerous projects concerning nanotechnology are supported by a number of specialist programs (e.g. WING – Innovative Materials for Industry and Society, IT research 2006, development program in optical technologies, supporting program in biotechnology).

BMBF funding in project development

Table 2.1 Expenditure on nanotechnology within the framework of different BMBF focal subjects (derived in 2003; estimation for 2004 and 2005).

Nanotechnology funding by the BMBF (in million EUR)	Priorities	1998	2002	2003	2004	2005
Nanomaterials	Nanoanalytics Nanobiotechnology Nanostructure materials Nanochemistry CCN Nano-competition for young professionals Nanochance		19.2	20.3	32.7	38.1
Production technologies	Ultrathin coatings Ultraprecise surfaces		0.2	0.8	2.2	2.2
Optical technologies	Nanooptics Ultraprecise processing Microscopy Photonic crystals Molecular electronics Diode laser OLED		18.5	25.2	26	26
Microsystems technology	Systems integration		7	7	9.4	10.2
Communication technologies	Quantum structure systems Photonic crystals		4.3	4	3.6	3.4
Nanoelectronics	EUVL, lithography, mask technology e-Biochips Magnetoelectronics SiGe-electronics		19.9	25	44.7	46.2
Nanobiotechnology	Manipulation techniques Functionalized nanoparticles Biochips		4.6	5.4	5	3.1
Analyses of innovations and techniques	ITA-Studies		0.2	0.5	0.2	
Total (in million EUR)		27.6	73.9	88.2	123.8	129.2

From 1998 to 2004 the funding volume of interconnected projects which concern nanotechnology has increased four-fold and currently amounts to approx. 120 million EUR. A list of the BMBF expenditure on nanotechnology research in different focal subjects for the fiscal years 1998 and 2002 to 2005 is given in Table 2.1.

2.6.1.2 Federal Ministry of Economics and Technology (BMWi)

BMWi

In addition to funding from the BMBF, the BMWi finances project-related investments in the Physikalisch-Technischen Bundesanstalt (PTB) and the Federal Institute for Materials Research and Control as well as nanotechnology-related projects in the program "Innovation Competence PRO INNO" for SME. For this purpose, approx. 25 million EUR are provided each year.

2.6.1.3 Institutional Research Funding

Other public expenditures in Germany

In Germany institutional research on nanotechnology outside the universities is in the hands of the four major research groups: the WGL, HGF, FhG and MPG. They run numerous research facilities or teams that have nano research as a focal topic. Moreover, these partners are involved in many special fields of research and programs on focal topics of the DFG. Table 2.2 shows the public expenditure of the DFG project development and of the institutional funding of the BMBF together with the Länder for nanotechnology-related research.

Table 2.2 Financial means for nanotechnology research within the framework of institutional funding and funding by the DFG.

Institutional funding of nanotechnology	2002	2003	2004	2005
DFG (Deutsche Forschungsgemeinschaft)	60	60	60	60
WGL (Wissensgemeinschaft G.W. Leibniz)	23.7	23.6	23.4	23.5
HGF (Helmholtz-Gemeinschaft)	38.2	37.1	37.4	37.8
MPG (Max-Planck-Gesellschaft)	14.8	14.8	14.8	14.8
FhG (Fraunhofer-Gesellschaft)	4.6	5.4	5.2	4.9
Caesar	1.8	3.3	4	4.4
Total (in million EUR)	**143.1**	**144.2**	**144.8**	**145.4**

2.6.1.4 Public R&D Funding in Germany

Table 2.3 shows an overview of the total amount of funding for nanotechnology in the years 2002 to 2005 in Germany.

Without the additional contributions of industry to project funding, the German expenditure relating to the public financing of nanotech-

Table 2.3 Expenditure of the public sector for the funding of nanotechnology projects in Germany 2002–2005.

Funding of nanotechnology in Germany	2002	2003	2004	2005
BMBF Project Funding	73.9	88.2	123.8	129.2
BMWi Project Funding	21.1	24.5	24.5	23.7
Institutional Funding	143.1	144.2	144.8	145.4
Total (in million EUR)	238.1	256.9	293.1	298.3

nology amounts to a total of about 300 million EUR for the year 2005. This volume excludes both the expenditure of the Länder, which contribute the basic financing for universities, and the financial resources allocated to nanotechnology research by industry *apart from* public financing.

2.6.1.5 **Assessment of the Situation in Germany**

In the field of nanotechnology, Germany can build on a history of well-trained scientists, a differentiated and linked landscape of R&D and institutes, as well as committed engineers and entrepreneurs. All participants are aware of the fact that although nanotechnological innovations require high investment, they create new job opportunities, too. The Ministry for Education and Research supports such innovative enterprises within the framework of the funding of combined projects, particularly in fields of application where a dominant market position and the envisaged realization of high margins appear feasible. Both future-oriented companies and the public sector commit considerable funds to the strengthening of this topic and its players. Both R&D work and the consolidation of flanking measures are addressed, such as the establishment of network structures, the organization of university courses for nanotechnology and other work developing young talents, as well as the integration of society into this topic.

2.6.2 Comparison with International Activities

Nanotechnology in Germany in the international comparison

Germany is among the world's leading countries in most fields of technology in the development of bases for new products and applications, and even in the field of nanotechnological R&D, Germany is considered to be on a level comparable to the USA and Japan. A general comparison of the distribution of publications and patents across the various countries shows, however, that in Germany the scientific domains of nanotechnology are still worked on separately from the application-related and product-related R&D areas (compare Hullmann, 2001). This situation is more or less comparable to the developments in Japan, whereas the USA is more closely geared towards realization-oriented aims.

Consequently, despite its quality as research location, its numerous business start-ups and its market perspectives, Germany has a good deal to make up concerning the realization of its nanotechnological know-how. With the establishment of competence centers, the funding possibilities for SME and educational work occurring at the same time, the Ministry for Education and Research has opened up an alternative funding approach to remedy this situation and to lay the foundations for future competitiveness. With the activation of a parallel funding strategy – project funding and the concurrent organization of supporting infrastructure – not only did Germany achieve a top international position in research in nanoscience, but even the companies that specialize in nanotechnological products have increased considerably, both in number and reputation. At a guess, the USA and Europe have nearly the same number of nanotechnology-related companies. About half of the companies domiciled in Europe are from Germany. A comparison with the situation in Japan or other countries in Southeast Asia is difficult, because reliable lists of companies do not exist for this continent.

Problem of comparability of expenses

However, it was difficult to compare the situation in other countries, both to assess industrial activities, and also to discover the relative expenditure:

- Depending on the definition of the technological field, some projects are included that would not fulfill the criteria of other countries.
- From the data recorded it is not clear whether full or marginal costs are presented. Furthermore, in the case of full costs, the proportions of overhead costs differ considerably. Sometimes, costs for the construction of buildings or the complete equipment of buildings are included.
- The differences in purchasing power in the individual countries result in different possibilities of developing programs. In countries with low labor costs (such as China), it is possible to invest many times the man-years despite significantly lower funding expenses.
- The proportion of industry expenditures can be specified only with difficulty.

Therefore, in Table 2.4 only the figures concerning public expenditures in the USA and Japan are given, without a detailed depiction of the funding activities.

Table 2.4 Expenditure on funding nanotechnology in Germany, Europe, the USA and Japan in million EUR, status of 2003 (for the sake of simplicity it is assumed that 1 USD = 1 EUR = 100 YEN; the comparability of the information is doubtful, as there is no internationally unique definition of the field; the details differ concerning gross or net costs; differences in the purchasing power are not taken into consideration; and the expenditure by industry can be specified only with difficulty.)

Funding of nanotechnology (in million EUR)	2001	2002	2003	2004
Germany	210	240	250	290
Europe (incl. nat. funding)	360	480	700	740
USA	465	697	862	989
Japan	600	750	800	800

A comparison of expenditures in Europe, in the USA and in Japan – roughly estimated and without paying close attention to the funding details – results in similarly high funding volumes. In the financial year 2002, about 700 million USD were spent in the USA and in the year 2003, about 860 million USD are allocated[1] within the framework of the "National Nanotechnology Initiative". The "Governmental Budget Plan for Nanotechnology" of the Japanese shows funding expenditure of approx. 750 million EUR for the year 2002, and as of 2003 in the region of 800 million EUR will be available annually. Even the British Government has recently announced an initiative that will guarantee expenditure of approx. 130 million EUR for the next six years, starting with the year 2004. The General Research Management of the European Commission estimates a volume of approx. 700 million EUR for the total nanotechnology funding in Europe in 2003. With the funding activities of the 6th EU-Research Supporting Program (FP6) – which supports nanotechnology mainly as the 3rd topical priority – the European Commission itself has planned a volume of altogether 700 to 750 million EUR until 2006, i.e. about 250 million EUR annually as of 2003. According to conservative estimates, with likewise 250 million EUR of public funding per year, Germany has the largest share in the national funding of nanotechnology in Europe. Therefore, the total means of funding of 200 to 250 million EUR spent on nanotechnological R&D by the other member countries can be deemed realistic.

1) For 2005 the USA is planning another considerable increase in the framework of the National Nanotechnology Initiative. Then, approx. one billion USD might be allocated.

Currently, all relevant industrial nations have recognized nanotechnology as an important future field and have promoted it accordingly. Therefore, national programs or those specific to research areas have been initiated not only in the USA, in Japan and Europe, but also in China, Korea, Taiwan, Australia and in other countries. Apart from high investment in this future field of technology, further characteristics of the current funding measures, already addressed by BMBF since 1998, are apparent in nearly all of these countries:

General high-tech aims of nanotechnology support

- an interdisciplinary approach to support the field
- concurrent funding of basic research and applied research
- initiation of network activities
- consolidation of international cooperation
- combination with problems of training and further education in future
- public discussion about society-relevant topics
- the urge for fast realization of concepts to strengthen the location.

2.6.2.1 Conclusions on the International Situation of Germany

The visibility of German activities has been increased recently by the nanotechnology-related measures taken by the BMBF. According to a statement of Philippe Busquin (former Research Commissioner of the EU), "Germany is the growth-engine in the EU in the field of nanotechnological innovations", (VDI-News, January 9, 2004). This shows that Germany's position in nanotechnology is good, both from the technical and infrastructural points of view. Earlier than other nations, the BMBF initiated the development of this future technology and addressed the wide range of this topic rightly in several subject programs. because of the focus on industry-oriented problems, certain sections reached an internationally significant position that must now be consolidated by adequate steps. To be able to compare well with the growing international competition, the innovation process must be accelerated and the long-term added-value must be ensured by innovation-accompanying measures.

Germany's position in nanotechnology is good

3
Application and Market Prospects of Nanotechnology in Products and Product Groups

Wolfgang Luther, Andreas Hoffknecht, Dirk Holtmanspötter and Volker Wagner

3.1
Notes on Literature Analysis

The analysis of application and market prospects of nanotechnology in products and product groups described here is mainly based on a secondary analysis of relevant publications. The following sources have been used:

- market analyses and press releases of different market research institutes, partly with specialization in the field of nanotechnology, e.g. Business Communication Company (BCC), CMP Científica
- press releases and publications of enterprises
- publications in specialist journals, the daily press and on the internet.

Quantitative evaluation of the market potential of nanotechnology is problematic

Apart from a qualitative description of the application potential, quantitative statements concerning the market potential of nanotechnology are also made here, albeit with the proviso that the field of nanotechnology shows the following characteristics that complicate a quantitative evaluation of the market potential:

- Nanotechnology cannot be assigned to classical lines of industry where the turnovers are recorded in business statistics. It is rather a cross-sectional discipline of different sectors of industry.
- Nanotechnology does not represent a uniform technological platform, but comprises a wide spectrum of various fields of technology and research (materials technology, coating technology, nanostructuring, analysis and surface treatment).
- Nanotechnological processes and products predominantly start at the beginning of the value-added chain and mainly refer to individual components only, whose functionality is enhanced by nanotechnology. The proportion of nanotechnology in the value added to marketable products cannot easily be assessed and can be recorded only vaguely.

Growth Market Nanotechnology. Edited by N. Malanowski, T. Heimer, W. Luther, and M. Werner

ISBN: 3-527-31457-1

- For the most part, nanotechnology is still in the research stage. Evaluations of the market success of future product options and predictions are therefore more or less speculative.

Market figures concerning nanotechnology show considerable discrepancies

The span of existing market figures concerning nanotechnology shows a wide range from 900 million USD for the world market volume of nanostructured materials in the year 2005 (BCC, Business Communication Company 2001) up to one billion USD for the world market volume of nanotechnologically influenced products in the year 2015 (NSF, National Science Foundation, 2001).

Figure 3.1 shows a logarithmic plot of different market forecasts with respect to the reference year. This shows that, up to the year 2015, is possible (to a rough approximation) to extrapolate the exponential growth of the world market for nanotechnological products. However, the different market forecasts are partly based on rather variable def-

Table 3.1 Outline of market figures and forecasts for nanotechnology.

World market (year)	With regard to	Source
493 million USD (2000) 900 million USD (2005)	Inorganic nanoparticles and powders (metal oxides like SiO_2, TiO_2, metals etc.)	Rittner (2002)
40 billion USD (2002)	Synthetic nanoparticles as primary products	Distler (2002)
23 billion USD (2003) 73 billion USD (2003)	Nanomaterials Nanotools Nanodevices + Nanobiotec	Fecht et al. (2003)
7 billion USD (2002) 28.7 billion USD (2008)	Nanotechnological products (mainly nanomaterials, also nanodevices and tools)	BCC (2004)
54 billion EUR (2001) 100 billion EUR (2005)	Nanotechnological products (classified according to nanomaterials, nanolayers nanoanalysis, ultraprecise surface treatment, lateral nanostructures)	Bachmann (2001), DG/GZ Bank (2001)
66 billion USD (2005) 148 billion USD (2010)	Nanotechnological products	Kamei (2002)
Up to 200 billion EUR (2005)	Nanotechnological products	Sal. Oppenheim (2001)
225 billion USD (2005) 700 billion USD (2008)	Nanotechnological products	NanoBusiness Alliance (2001)
1 billion USD (2015)	Nanotechnological products	NSF (2001)

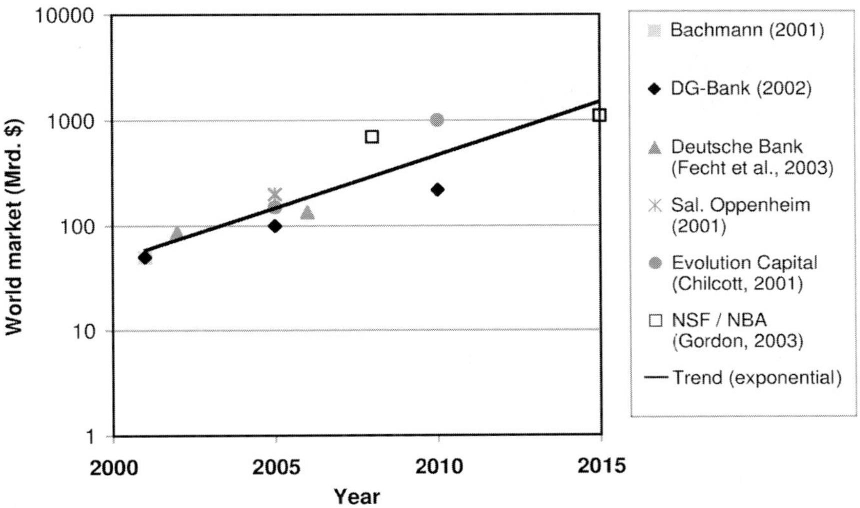

Figure 3.1 Comparison of different market forecasts for the world market for nanotechnology. (Source: Bachmann, 2001; Fecht et al., 2003; DG-Bank 2002; Gordon, 2003; Sal. Oppenheim, 2001; Chilcott et al., 2001).

initions and approaches of evaluation, and therefore should not be considered as directly comparable.

In order to evaluate the market figures and forecasts correctly and to avoid misinterpretations we need to consider the following questions:

- Which definition of nanotechnology forms the basis?
- Which databases are the source of the stated figures?
- Which (sub)areas of nanotechnology does the market evaluation comprise (e.g. nanomaterials, nanocoatings, tools and measurement technology for the generation of nanostructures etc.)?
- Which value-added stage do the market figures refer to (elements such as nanopowder, intermediate products like laser diodes or end user products like computers or domestic appliances)?

Different evaluation approaches for market forecasts for nanotechnology

The market forecasts of BCC and NSF mentioned above take two extreme views for their approach to evaluation. While BCC keeps to a closely limited section of nanotechnology (inorganic nanoparticles) calculating the market value of the elements, NSF refers to the market value of all end products that are somehow influenced by nanotechnology (e.g. medicine, computer, data storage etc.), without mentioning specific products and without specifying the proportion of nanotechnology in the added value. Therefore, it is hardly surprising that the given market volumes differ by a factor of 1000, in which the different time horizon certainly plays an important role.

Segmentation in different sub-areas of nanotechnology

Between these two extremes, other market forecasts segment nanotechnology into different sub-areas (nanomaterials, nanoanalysis, etc.) and determine the market potential of nanotechnology by adding up the market value of specific products that contain nanotechnological components (e.g. Bachmann, 2001, DG Bank, 2002, Fecht et al., 2003). For products that have not yet reached market maturity, the substitution potential for existing products is usually given (e.g. MRAM-memory chips as substitution for DRAM-memory chips). If the value-added proportion of a nanotechnological component in the end product is not quantifiable or market data is not available, the value of "the smallest saleable unit" of a product is stated that contains the nanotechnological function (e.g. in the case of hard disk data storage, the whole disk drive, although only the GMR reading head contains a nanotechnological function).

Moreover, the market figures may be distorted by double references being made in the evaluation of nanoproducts. So it may be that products of different value-added stages are repeatedly brought into the evaluation, even though they are based on the same nanotechnological basic product. Therefore, it is possible that, for example, to determine the total market volume, a nanocrystalline material in a product (e.g. suncream) is brought in both on the value-added stage of the raw material (nanocrystalline material) and of the end product (suncream). These double references could lead to an overestimation of the market volume.

For this reason the following analysis attempts to depict the application and market potential of nanotechnology in the context of the value-added stage examined and the value-added proportion of the nanotechnological component as well as of the time horizon. The analysis was carried out both segmented according to the relevant technological sections of nanotechnology (Chapter 3) and broken down according to the different important branches of economy (Chapter 6).

3.2
Nanomaterials

Nanoscale materials are the essential basis of the whole field of nanotechnology. They show many extraordinary properties not found in conventional materials. On the structural side, these include e.g. super-elasticity, increased hardness, fracture toughness and fracture strength. On the functional side, these properties include e.g. improved soft-magnetic qualities, giant magnetoresistance effect, re-

Table 3.2 Examples for adjustable properties of nanomaterials.

Property	Examples of effects caused by nanoscale configuration
catalytic	Increased catalytic effects through considerably enlarged surface
electric	Increased electrical conductivity in ceramics and magnetic nanocomposites, higher electrical resistance in metals
magnetic	Increased magnetic coercivity up to a critical grain size (below this size, reduction of coercivity up to super-paramagnetic reaction)
mechanical	Increased hardness and strength of metals and alloys, improved ductility, hardness and malleability of ceramics
optical	Spectral displacement of the optical absorption and fluorescence properties, increase in luminescence of semiconductor-crystallites
steric	Increased selectivity and effectiveness of membranes, adaptation of cavities for the delivery or the controlled discharge of specific molecules
biological	Increased permeability for physiological barriers (membrane, blood-brain-barrier etc.), increased bio-compatibility

New effects through material design on the nanoscale

duced thermal conductivity or higher electrical resistance. Nanoscale materials are usually manufactured in well-known processes. They range from inorganic and organic, amorphous or crystalline nanoparticles that may occur singularly in aggregations or as powder or even dispersed in a matrix, through nanocolloids, nanosuspensions and nanoemulsions, to the family of fullerenes and their derivatives. Table 3.2 outlines the essential physicochemical and biological properties of nanomaterials that allow the specific adjustment and optimization owing to their nanoscalability.

Essential properties of nanomaterials in contrast to classical macroscopic substances are their large ratio of surface, or interface, to volume as well as the quantum effects which are prevalent. For a particle with a diameter of 10 nm, for example, about 20% of all atoms are on the surface; in the case of a particle with a diameter of 1 nm it may be more than 90%. With an increasing surface share, the surface energy of the single particles increases too, causing, for example, the melting point to reduce or the sinter activity to increase. Through precise control of the sizes of the particles, their properties can be adjusted to a certain extent. Mostly, it turns out to be difficult to maintain these desired properties beyond the ensuing manufacturing process. So loose quantities of nanopowders of a large number of materials tend to grow together to form larger particles or firmly combined agglomerates through diffusion processes even at room temperature. Therefore, a suitable manufacturing process or secondary "refining process" has to be chosen or developed even for the manufacture of nanoscale basic products, in order to impede or prevent agglomerates and grain growth before and during the manufacture and operation of the workpiece.

Solid materials can be produced from the vapor phase, the liquid phase and from solids in a way that they are nanoscale (i.e. by definition smaller than 100 nm) in at least one dimension. These products exist mostly in the form of particles or thin layers. Especially with methods using the vapor or liquid phase for the manufacture of the particles or layers, self-organization processes that develop certain forms and structures are frequently used.

Wide range of methods for the manufacture of nanoparticles and nanocoatings

Several relevant methods are known for material syntheses from the vapor phase. They are Chemical Vapour Deposition (CVD), Physical Vapor Deposition (PVD), aerosol-based methods such as Chemical Vapor Condensation (CVC), sputter methods and flame synthesis. While the latter method is only suitable for the production of powders, with the others the deposition of both powder and thin coatings is possible. There are different variants of each method that can often be applied continuously, a feature that makes them particularly interesting for industrial processes.

The most important methods for the manufacture or deposition of powders and thin coatings from the liquid phase are the sol–gel process and electrochemical deposition. Furthermore, both processes are suitable for the development of nanoporous volume objects. Nanoporous solids may be realized by polymer–pyrolytic methods, too. Apart from this, for the production of powders, sonochemistry (ultrasonic chemistry) and hydrothermal methods are particularly important. Moreover, ultrathin surface films can be developed by laser-beam melting.

The manufacture of powdery nanomaterials from the solid phase mostly occurs by means of high-energy or high-speed grinding as well as by irradiation with particles, e.g. ions, and ensuing recrystallization. The grinding processes are technologically undemanding; however, they involve the danger of strong contamination through abrasion. In addition, there are more exotic methods, such as the explosive sputtering of metallic wires. Another very important method is the nondestructive *in situ* generation of nanostructures in volume objects. This is primarily about the controlled crystallization of amorphous material to nanocrystalline composites. This method is already state of the art for oxide glasses, and is gaining in importance for metals and alloys. Nanoporous solids for example, can be developed through electrochemical oxidation processes from dense metallic volume objects.

3.2.1
Classification of Nanomaterials

All classical materials such as metals, semiconductors, glasses, ceramics or polymers may be available in nanoscale configurations, i.e. they can be produced in particle sizes smaller than 100 nm. Meanwhile, supramolecular structures, products of a research trend in chemistry for a long time (e.g. dendrimers, micells or liposomes), are also classed as nanomaterials just like Langmuir–Blodgett films, for instance, or more up-to-date material classes such as fullerenes. In principle, there are several possible ways of classifying nanomaterials; some variants are shown in Table 3.3.

Different approaches for the classification of nanomaterials

To achieve efficient application of nanomaterials, factors relating to their use in nanoproducts may serve to distinguish classes (Haas et al., 2003).

In the following, the application and market potentials of classes of nanomaterials will be discussed that are particularly economically relevant in the value-added stage of raw materials and intermediate products.

Table 3.3 Examples for classification possibilities of nanomaterials.

align items vertically in right hand column as examples of bullet point in left hand column

Classification according to	Examples
Dimensions	
• 3 dimensions < 100nm	Particles, hollow spheres...
• 2 dimensions < 100nm	tubes, fibers, wires ...
• 1 dimension < 100nm	Films, layers, multilayer ...
Phase composition	
• Single-phase solids	crystallines, amorphous particles and layers...
• multi-phase solids	matrix materials, coated particles...
• multi-phase system	colloids, aerogels, zeolites...
Manufacturing process	
• vapor-phase reaction	Flame systems, condensation, chemical vapor deposition...
• liquid-phase reaction	sol-gel, precipitation, hydro-thermal process ...
• mechanical processes	sphere grinding, three-dimensional deformation ...

Table 3.4 Examples for classification possibilities of nanoproducts.

Classification according to	Examples
Value added stage	
• raw materials	layer silicates, nanopowder, precursors
• intermediate products	paints, adhesives, nanocomposites...
• semi-finished products	coated sheet blank, fabric material...
• components	sensor, electrode, Laser, LED ...
• systems/end products	fuel cells, notebook, bio-chip ...
Maturity/production volume	
• engineering sample	components for molecular electronics
• prototype	
• pilot application	switchable adhesives, multifunct. textiles...
	fluorescent markers, contrast medium, CNT,
• mass product	carbon black, aerosil, titanium dioxide
Processability	
• intrinsic nanomaterials	nanoproperties remain preserved in the product (quantum dots, fluorescent markers...)
• process nanomaterials	nanoproperties no more detectable in the product (e.g. agglomerated carbon black)

3.2.2 Nanoparticles/Fibers/Tubes

3.2.2.1 Metal oxides/Metals

Metal oxide nanoparticles are applied in a multitude of products

In the field of inorganic nanoparticles, metal oxides, especially silicon dioxide, ceroxide, titanium dioxide, aluminum oxide are currently of utmost economic importance. The main fields of application of these nanoparticles are electronics, pharmaceutics/medicine/cosmetics as well as chemistry/catalysis. In the sector of pharmaceutics/medicine/cosmetics the economically most important application is presently the field of sun protection products. Here, in particular nanoscale titanium dioxide particles, but also zinc oxide particles are applied as UV absorbers that, because of their nanoscalability, offer the advantage of optical transparency. Further fields of application for nanoparticles in medicine are as marker materials for biological quick tests (e.g. gold or semiconductors), contrast media for magnetic resonance tomography and antimicrobial coatings and composite materials for sterile surfaces and medical appliances. From the economic point of view nanoparticle-based drug carrier systems are currently still insignificant, but promising on the long run are for the specific and selective application of active substances.

In the field of catalysis, nanoparticles that are used as a porous backing layer for catalytic converters of cars make up the major mar-

Table 3.5 Applications and market potentials of nanoparticles (acc. to Rittner, 2002).

Fields of application/materials	World market volume 2005
Electronics, optoelectronics • Chemical-mechanical polishing (e.g. Si, Al-Oxide) • Electrically conductive coatings (e.g.. ITO, precious metals) • Magnetic data storage (ferric oxides, cobalt oxides) • Condensers (barium-strontium-titanium, composites)	668 million USD
Pharmaceutics/medicine/cosmetics • Sun protection (titanium dioxide, zinc oxide) • Biomarker (gold, semiconductors, ...) • Contrast medium (ferric oxide, chelate complexes...) • Biomagnetic separation (ferric oxide) • Antimicrobial agents (silver, titanium dioxide ...)	145 million USD
Chemistry/catalysis • Catalytic converters for cars (aluminum oxide, precious metals ...) • Structural ceramics (oxide ceramics, SiC ...) • Scratch-resistant coatings (different metal oxides and carbides) • Photocatalysis (titanium dioxide) • Fuel cells (precious metals, YSZ) • Ceramic membranes (different metal oxides)	88 million USD

ket share. Here, mainly nanoscale aluminum oxide is used which serves as a porous carrier for the precious metal catalyst, which is finely spread in situ and deposited on the substrate. The importance of nanomaterials for catalytic converters in hydrogen reformers for PEM-fuel cells will increase as well.

BCC estimates the market volume for oxide and metal nanoparticles at 900 million USD in the year 2005. Table 3.5 gives an overview of applications and market potentials.

3.2.2.2 Carbon Particles

Carbon nanoparticles, carbon black, fullerenes and carbon nanotubes

Nanostructured carbon particles comprise both established materials produced on a large-scale industrial basis, such as industrial carbon black, and relatively young material classes, such as fullerenes and fullerene derivates. Special material classes within the fullerenes are the single-wall and multi-wall carbon nanotubes that are forecast to have a high product potential on a medium and long-term basis.

The classical nanostructured materials such as carbon black and special blacks are of great economic significance, and presently dominant with a current world market volume ranging between 3 billion

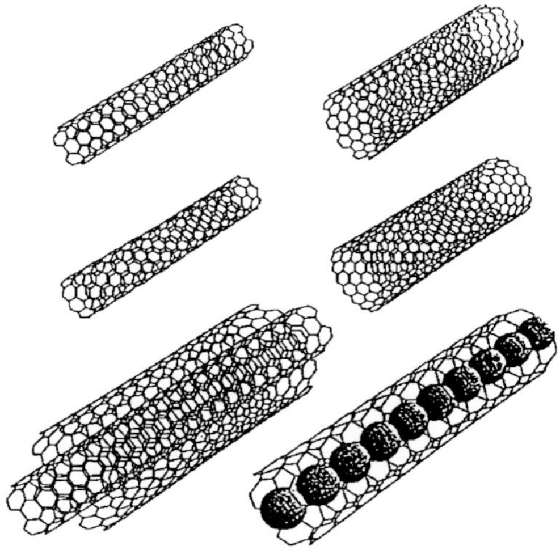

Figure 3.2 Different modifications of carbon nanotubes (single-wall, multi-wall, filled with dopants etc.).

USD (Reuters, 2003) and 5.7 billion USD (SRI, 2002) or a forecast world market volume of 8 billion USD (Fecht et al., 2003) in the year 2006. Carbon black consists of chain-like, agglomerated carbon particles with their primary particle size lying in the nanometer range. The main fields of application of these materials, that are synthesized by flame aerosol processes, are fillers for rubber and pigments, for example for car tires or toners.

A high economic potential is predicted for carbon nanotubes (CNT) in future becuse they have extraordinary molecular properties. They show extremely high tensile strength: on a molecular basis a tensile strength approx. 100 times higher than steel with a specific weight being 6 times lower, (CMP Cientifica, 2003) as well as excellent thermal and electrical conductivity. Extensive application of carbon nanotubes could be made, e.g. in sensor technology, in electronics (CNT-based connecting conduits and transistors), in composite materials (e.g. electrically conductive polymers) or in flat screens (electron emitters in field emission displays). However, an economic obstacle is the high price of about 180 USD per gram (for single-wall CNT) (Löfken und Mayr, 2003), which is the result of the still unsolved problem of the large-scale manufacture of carbon nanotubes with defined composition. While the present market potential of CNT may lie at a few million US dollars, very optimistic forecasts already predict a market volume of 1.2 billion USD by 2006 (Fecht et al., 2003). These forecasts are based on the assumption that inexpensive mass production of CNT will be established on a short-term basis, as has already been announced by a Japanese company, for instance (Mitsui, 2002).

3.2.2.3 **Layer Silicates**

Nanostructured, organically modified layer silicates (nanotones) have been used for a while as fillers for polymer materials for the improvement of the barrier properties (e.g. gas tightness), as fire protection and for mechanical reinforcement. Although some products are already on the market, problems in the manufacturing process as well as relatively high costs with moderate performance gains prevent the extensive economic application of these materials. Until 2006, the world market volume for nanolayer silicates is assessed at 25 million USD (SRI, 2002).

3.2.2.4 **Organic Nanoparticles**

Among others, the following material classes are included under the group of organic nanoparticles (comp. Horn and Rieger, 2002):

- Polymer nanoparticles/dispersions
- Micronized active agents and effect substances (vitamins, pigments and pharmaceutical products)
- Macromolecule (e.g. dendrimers)
- Micelles, liposome

High market potential for micronized agents and polymer dispersions

From today's point of view, mainly micronized drugs and substances and polymer dispersions are economically relevant. Organically active agents and drugs, such as vitamins, pigments and pharmaceutical products are often poorly soluble in water and require special formulation procedures for applications in aqueous form; they can be micronized, which improves the surface-to-volume ratio, and thus optimizes the water solubility and with this the physiological (pharmaceutics, cosmetics, pest control, nutrition) or technological effectiveness (lacquers and printing colors). Such nanoparticles can be manufactured by mechanical size reduction, precipitation or condensation from colloidal solutions. The market potential for micronized substances (especially vitamins) is assessed at approx. 1 billion EUR for the year 2002 (BASF/Distler, 2002). With approx. 15 billion EUR in the year 2002 (Distler, 2002), aqueous polymer dispersions show even higher market volumes. This material class, long established in industry can be optimized with regard to their technical properties by the application of modern nanotechnological procedures, for example, through the increase of the solids content due to controllable particle size, through selective surface modification of the polymers or the generation of nanocomposites through mixing with additional, even inorganic fillers. Such polymer dispersions offer wide fields of application, e.g. as binding agents in dyes and paints, adhesives for labels and sealing tapes or as coating systems for textiles, wood or leather. Moreover, aqueous polymer dispersions are environmentally friendlier than products based on organic solvents.

Organic macromolecules, such as dendrimers or similar substance classes like hyper-branched polymers (e.g. based on polyurethane) are still insignificant from an economic point of view, but are considered to have good market prospects in future (Bruchmann, 2002). Application potentials of dendritic molecules exist, for example, as carrier media for catalytic converters or pharmaceutical active agents (drug delivery) or even as cross-linking agents for scratch-proof car paint or printing colors. The global market potential of dendrimers is assessed at 5–15 billion EURO for the year 2006 (SRI, 2002).

3.2.3 Nanocomposite Materials

3.2.3.1 Polymer-based Nanocomposites

Strong market growth forecast for the field of polymer nanocomposites

Apart from the particular systems already mentioned above, also nanostructured block copolymers and polymer materials doped with ceramic, metal or even semiconductor nanoparticles are included with the polymer-based nanomaterials. The doping serves, inter alia, to improve (thermo-)mechanical, electronic or also biological qualities. The following examples can be cited:

- Nanosilicate-reinforced polymers for the improvement of barrier properties (e.g. gas tightness), as fire protection and for mechanical reinforcement.
- Epoxy resin filled with nanoparticles for application in car electrics as improved impregnating and casting resin for coils and windings
- Conductive polymers, e.g. through doping with carbon black or in future even with carbon nanotubes, for application in the electrostatic protection of electronic devices etc.
- Polymers filled with nanoparticles (e.g. of silver) with antimicrobial properties for application in medical technology.

Medium-term forecasts predict a very strong growth for the world market of polymer nanocomposites from 15 million USD in the year 2001 to 300 million USD in the year 2006 (SRI, 2002).

Another rapidly growing market segment is the organic semiconductors, which occasionally belong to the field of nanotechnology, since the molecular components lie in the order of some 10–100 nm and for the optimization of material properties, an understanding of the fundamental processes on the nanoscale is required. Underlying the recent rapid progress in the development of organic semiconductors was the synthesis of new types of substances, the improved purity of these materials, a controlled layer production as well as the effective protection of the substances and components against air and humidity. In contrast to conventional semiconductors, such as silicon or gallium arsenide, these polymers offer the advantage of simple manufacture and the possibility of processing them into large, mechanically flexible components. The market volume for organic semiconductors is assessed at 100 million EUR for the year 2006 (source: company survey) for applications mostly in OLED displays, in the long run even in solar cells.

3.2.3.2 Ceramic Matrix Materials

Application of synthetic nanopowder for the manufacture of ceramics

In the case of ceramics, the main focus is on the generation of nano/nano and micro/nanostructures. The aim is above all to enhance thermo-mechanical properties of this actually brittle material group, i.e. fracture toughness and the ability to thermoform ("superelasticity"). The application of those synthetic nanopowders is useful, that stand out due to their high chemical purity and adjustable powder grain sizes. As a rough distinction, sol–gel processes are preferably used for the generation of oxide powders (e.g. Al_2O_3, SiO_2), whereas for the production of non-oxide powders (e.g. Si_3N_4, SiC, TiCN) gas separating methods are mainly applied. Here, it is advantageous that with decreasing particle size, caused by increasing surface energy, the sintering temperature required for the consolidation and compression can diminish. Even nanostructured gradient materials – the gradient may be scaled with respect to (thermo-)mechanical, functional and chemical properties – are realizable in different ways; by electrochemical separation, separation processes from the vapor phase or reactive gas treatment.

Moreover, ceramic membranes and volume objects with very precisely defined porosity can be manufactured. They mainly serve selective liquid and gas separation, as particle filters, catalyst carriers and micro reactors. More modern low-temperature synthesis methods, such as the sol–gel method, are successful even to manufacture thermodynamically incompatible glass compounds. The manufacture of vitroceramics by defined crystallization, is possible for most different applications. In the "low-tech" field, this comprises mainly products of the household sector or armor in the safety field. In the "high-tech" sector, products from the field of implant and medical technology as well as laser and space technology are concerned.

3.2.3.3 Metal Matrix Materials

It is possible to manufacture thermally heavily loadable materials with high resistance through the reinforcement of metals with ceramic fibers, especially silicon carbide, but also aluminum oxide or aluminum nitride. Such Material-Matrix-Composites (MMC), e.g. SiC in Al-alloys or TiN in Ti/Al-alloys have a high application potential for structures in energy engineering (e.g. turbines) or even in aviation engineering and space technology, because of their high temperature stability, low density, high resistance, high thermal conductivity and the controllable thermal extension. Through the nanoscale structure of the Metal-Matrix-Composites, higher strength and resistance against material fatigue are achievable as well as better malleability and superelasticity. Particle-reinforced steels are also currently being developed, inter alia for applications in car manufacture. So by means of dispersion technology, nanometer-sized carbon nitrite particles can be applied to martensitic steel in order to optimize the fatigue performance of the steel (Masaki et al., 2003).

3.2.3.4 **Aerogels**

Aerogels, lightest solids with broad application potential

Aerogels are highly porous solids with an inner surface of between 600 and 1000 m^2 per gram. With a density between 0.003 and 0.35 g cm^{-3} they are regarded as the lightest solids known. Aerogels consist of nanoparticles arranged in a highly porous three-dimensional network. Apart from inorganic, mostly oxide aerogels, even the synthesis of organic polymers was carried out successfully some years ago. Aerogels display interesting properties for technical applications down to their nanostructure and high porosity. The outstanding technically exploitable properties of aerogels include:

- Extremely low thermal and acoustic conductivity for applications as insulating material in building engineering, process engineering, automotive engineering, textile engineering and further fields of application
- Optical transparency/translucency for applications in building engineering (heat-insulating window panes and fronts) and in sensor technology
- Very low dielectric constant for applications in electronics.
- Extremely high inner surfaces and porosity for controllable drug delivery for applications in the chemical and pharmaceutical industry, and for membranes and filters in energy and environmental technology.

The limits of economical large-scale production of aerogels have already been passed, so that commercial product options with a multitude of application fields are available. The Aspen Systems company assesses the world market potential for such economical aerogel materials at 10 billion USD by the year 2005 (Aspen Systems, 2001). Even if this figure is an exaggeration, since the commercial mass production of aerogels, e.g. through Cabot, is currently still in an early stage, trends indicate a high economic potential. Figure 3.3 gives an overview of product options and fields of application.

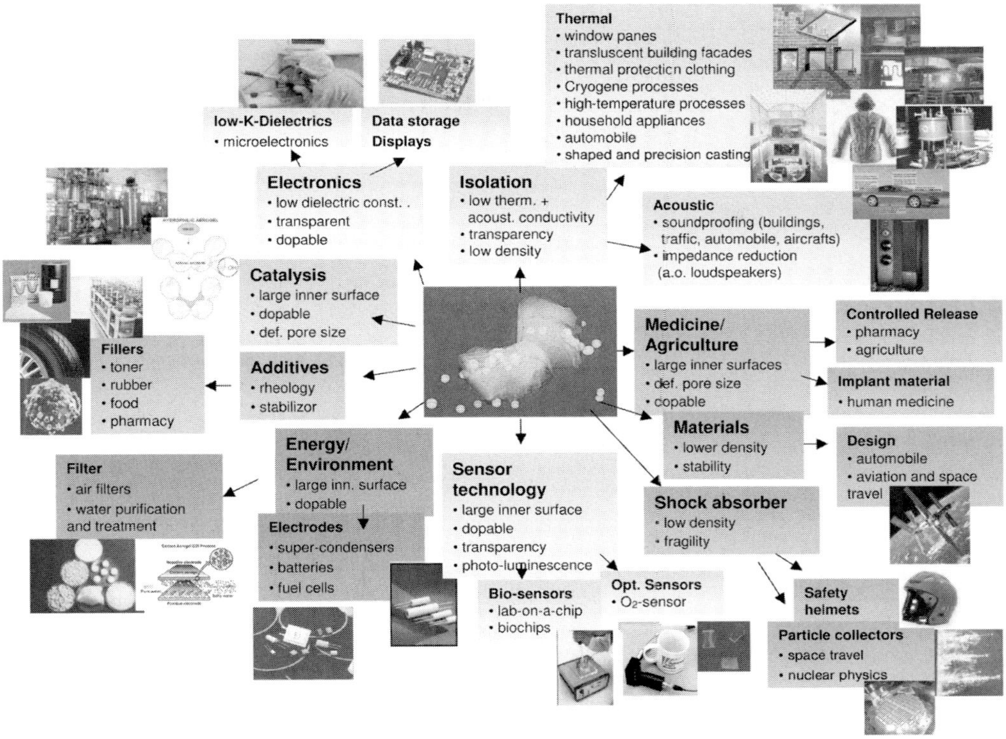

Figure 3.3 Product and application options of aerogels (Source VDI TZ).

3.2.3.5 **Zeolites**

Similar to aerogels, zeolites are highly porous solids on the basis of alumosilicates that occur in nature and are manufactured artificially as well. Owing to their large inner surface and their cage-like pores that allow the absorption of "guest molecules", they are used as ion exchangers, molecular sieves and catalytic converters. One main field of application for zeolites is their addition to detergents and cleansing agents to reduce the hardness of the water by bonding calcium and magnesium ions dissolved in the water. The world market volume for zeolites is stated to be 2.3 billion USD for the year 2002 with an increase of up to 2.6 billion USD by 2006 (Fecht et al., 2003).

3.2.4 **Nanocoating Systems**

Nanoscale coating systems represent a function-determining element in many product applications. A multitude of physical, chemical or biological effects are achieved by a nanoscale coating design that can be used for commercial products. Table 3.6 outlines properties that are

Table 3.6 Product properties adjustable by means of nanoscale layer systems.

Material properties	Layer types (examples)	Application examples
Mechanical properties (tribology, hardness, scratch resistance)	Rough matter layers of metal carbides and nitrides, diamond-like layers (DLC), sol-gel coatings	Wear protection for mechanical devices, e.g. computer hard disks, mechanical protection for soft materials (polymers, wood, textiles), e.g. scratch-proof plastic lenses for glasses
Wetting reaction (e.g. hydrophobia, hydrophilia)	Fluorinated hydrocarbons and fluoroalcylsilane	Anti-graffiti, antifouling, lotus-effect, self-cleaning surfaces for textiles and ceramics
Thermal and chemical properties (heat and corrosion resistance)	Metal oxides, carbides and nitrides, organic self assembled monolayers	Corrosion protection for metal surfaces (e.g. aluminum) heat protection for turbines and engines
Biological properties (biocompatible, antimicrobial)	Organic coatings, photocatalytic titanium dioxide coating	Biocompatible implants, anti-bacterial surfaces for medical devices and in the sanitary sector
Electronic, electric and magnetic properties (magnetoresistive, dielectric, conductive)	Metal multilayers indium-tin-oxide-coats silicon dioxide	Magnetoresistive sensors and data storage, dielectric layers for transistors, antistatic coatings for packaging and casing materials, transparent electrically conductive layers for solar cells and displays
Optical properties (antireflective, photochromic and electrochromic)	Nanoporous silicon dioxide layers for anti-reflecting, electrochromic tungsten trioxide layers	Photochromic and electrochromic glazing, antireflective panes and solar cells

adjustable due to nanoscale coating systems and gives examples for their product application (TAB, 2003).

Because there are so many different product classes and application fields in which nanoscale layer systems play a function-determining role, a complete record of the market volume is not possible at this point. In general, the proportion of function-holding nanoscale layer systems in the added value of the products is not quantifiable, as the coating or surface functionalization represents only a part of the production process, that is often carried out in a value-added stage of primary and intermediate products. Therefore, determination of the market value for the coating as such is not usually possible. Considering the value-added stage of end products, significant market volumes arise for products whose functionality is substantially determined by nanolayers. For the year 2001, the world market potential for such products has already been assessed at 21 billion USD (Bachmann, 2001). The market value of hard disk drives with GMR reading heads alone lies in the double-digit billion US dollar range (see Chapter 6.7). In the company survey, the following product groups based on nanoscale layer systems were identified that will have economic relevance for German companies in the period of time until 2006:

- Hard layers (world market volume in the year 2006 about 0.5–1 billion EUR)
- Tribological layers (world market volume in the year 2006 about 1–5 billion EUR)
- Antifog layers (world market volume in the year 2006 a bout 50–250 million EUR)
- Antireflex layers on plastic surfaces (world market volume in the year 2006 about 100 million EUR)
- Tool coatings (world market volume in the year 2006 about 50–250 million EUR)
- Corrosion protection coatings (world market volume in the year 2006 about. 1–5 billion EUR)
- Electronics based on functional nanolayers, e.g. GMR-HDD, MRAM (world market volume in the year 2006 >5 billion EUR)

3.3 Nanoelectronics

In this connection, nanoelectronics is interpreted as electronics based on silicon with structure widths below 100 nm and alternative approaches also based on nanotechnology. Furthermore, it includes process engineering and/or analytical equipment required for the manufacture of the components mentioned.

This definition orientates itself towards the BMBF funding program IT 2006 (BMBF, 2002) and describes the current transition of micro to nanoelectronics. For the future it is expected that new effects due to miniaturization will be increasingly exploited in electronic components. As shown in more detail in the following sections, it is exactly the aim of many research efforts in nanoelectronics to maintain the proven functioning of the CMOS electronics despite the shrinking dimensions.[1] In contrast to this, a lot of alternative approaches are aiming to exploit new effects (see Section 3.3.3). However, these approaches are feasible on a medium or long-term basis, and from an economic point of view they will not play a role in this survey within the timeframe until 2006 (Hoffknecht, 2003).

1) CMOS electronics is usually used as a synonym for the current mainstream-electronics on silicon basis. In the original meaning of the word, CMOS (Complementary Metal Oxide Semiconductor) stands for a logic based on complementary n-channel or p-channel MOS field effect transistors.

The International Technology Roadmap for Semiconductors (ITRS) classifies the electronics generations according to half the distance of the electronic lines of a DRAM.[2] This parameter is called DRAM 1/2 Pitch. In 2003, DRAM were merchandized for the first time ever with the DRAM 1/2 Pitch reaching the 100 nm mark. The present survey uses this development as the entry into nanoelectronics, as this corresponds to international customs and is therefore particularly practicable for the investigation of the market potentials.

Without starting a more academic discussion, it has to be pointed out that there are other criteria to be considered:

- The gate length of transistors, for example, uses smaller dimensions than the DRAM 1/2 Pitch. In 2003, the lithographically generated gate length of transistors was already 53 nm (Fig. 3.2). Using special etching techniques, this length was reduced to only 37 nm in a further production process. With the 180 nm generation introduced in 1999, the actual gate length of transistors had already fallen below the 100 nm mark (ITRS, 2003).
- While the lateral structures are currently at the 100 nm threshold, the components have been using vertical coats with thicknesses of below 100 nm for a long time. In many cases these small layer thicknesses are necessary for the functionality of the components.

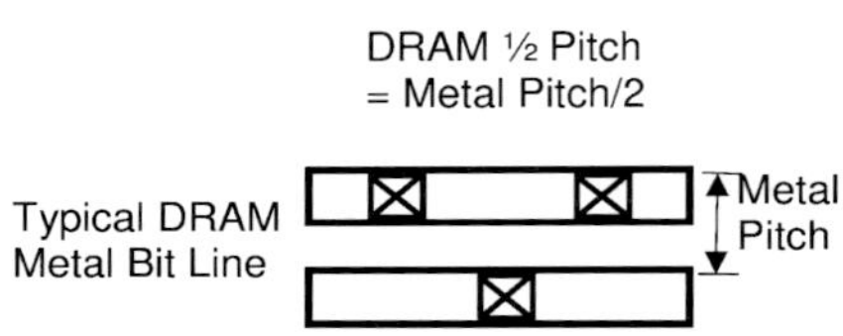

Figure 3.4 Definition of the DRAM 1/2 pitch as half the distance of two DRAM lines (Source: ITRS, 2003).

3.3.1
Si-based Electronics

The success story of electronics is closely connected to the progress of silicon technology.

In 1965, Gordon E. Moore – later co-founder of the company Intel – was questioned by the *Electronics* magazine about the further development of semiconductor electronics. He formulated an empiric law – six years after the invention of the Integrated Circuit – that was later called Moore's law (Moore, 1965; Mann, 2000). According to Moore's law, the performance of an IC and the number of components crammed on a chip double every 18–24 months – with the price per chip remaining constant.

Moore's Law

2) DRAM (Dynamic Random Access Memory): Semiconductor memories with the highest market share, in which digital information is coded by charges on condensers.

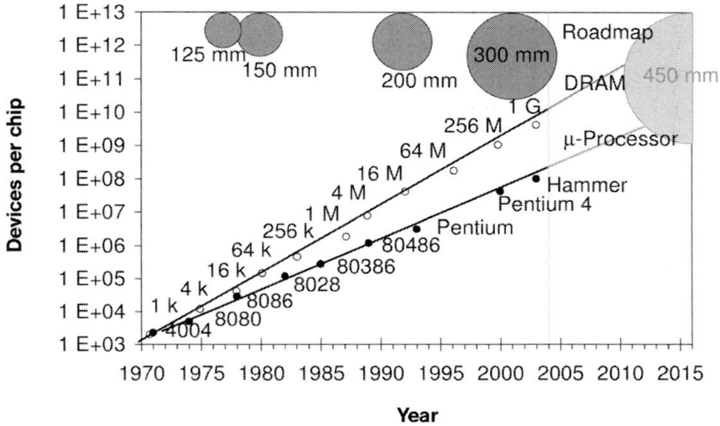

Figure 3.5 Development of silicon-based electronics. The number of components per chip in microprocessors and DRAM since 1970 is shown, and the ITRS forecast for the further development. In addition, the dimensional development of the Si-wafers used in production is marked. (Sources: ITRS, 2001; ITRS, 2002; ITRS, 2003; Normile, 2001; Intel).

Figure 3.5 shows the development of the number of components on a chip for microprocessors and for DRAM since 1970. This exponential growth is accompanied by an exponential decrease in costs per digital function by currently 29% per annum (see Fig. 3.6).

Success through miniaturization

This development is possible thanks to the continuously progressing miniaturization.

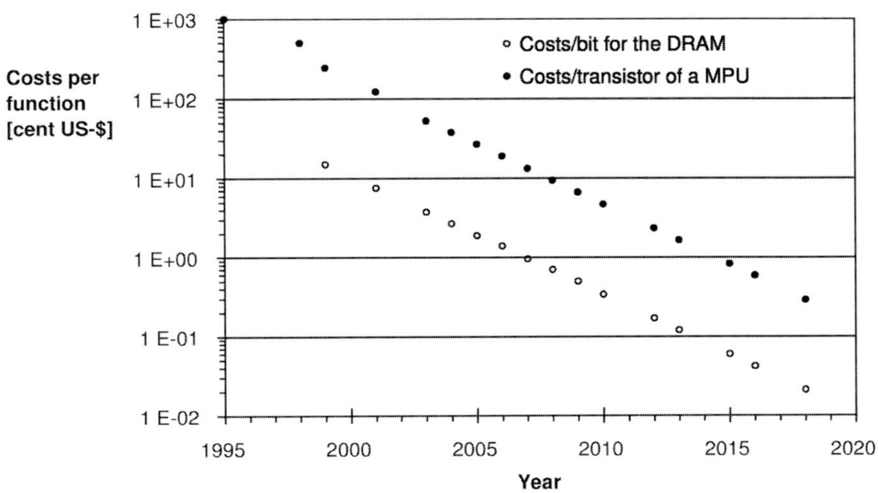

Figure 3.6 Development of costs per transistor and of the costs per bit for DRAM since 1970 in cent USD (Source: SIA, 1995; ITRS, 1999; ITRS, 2003).

Figure 3.6 shows the development of the DRAM 1/2 Pitch and the transistor gate length since 1999 and the forecast of the ITRS until 2018. CMOS technology stands out due to its particularly good-natured scalability, the principles of which have already been formulated in the 1970s (Dennard, 1972; Critchlow, 1999). The reduction of all measurements of a transistor in all dimensions by the factor α leads to a reduction in voltage, electricity and the switching time by α. Since the package density increases by α^2 while the power consumption decreases by α^2, the power dissipation per chip surface remains constant.

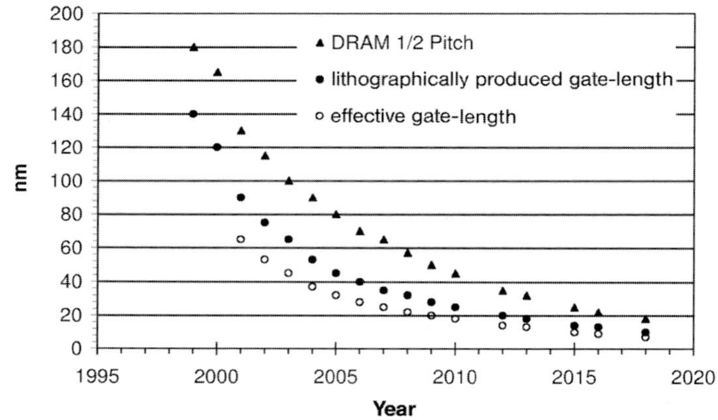

Figure 3.7 Development of the lateral dimensions of CMOS electronics (ITRS, 1999; ITRS, 2001; ITRS, 2003).

Despite this good-natured behavior, time and again technology meets with technological hurdles it has to overcome for a further development according to Moore's Law. Since 1998, the ITRS has been describing the further development of technology for the respective 15 years to come in yearly issues. They described these technological hurdles as so-called Roadblocks. Here, a distinction is made between Roadblocks whose solutions are principally known and already in the stage of development, and those for which realizable approaches are not yet known. Firstly the number of Roadblocks increases from issue to issue of the roadmap, and secondly the further they look into the future, the more the number of Roadblocks with still unknown solutions increases. This is depicted in Fig. 3.8 for the roadmap issues of 1999, 2001 and 2003. The figures are indicated for a period of 4, 9 and 15 years after the publication of the roadmap. This means that for the ITRS of 1999, the figures are given for the years 2003, 2008 and 2013. These facts are often quoted as an argument for the approaching end of CMOS technology (Normile, 2001).

Roadblocks

A more precise analysis of the roadmaps shows that these conclusions cannot be verified, since the relative proportion of the roadblocks in the increasing total number of the criteria considered in the roadmap diminishes from issue to issue (see Fig. 3.9). The fact that the total number increases is caused by the increasing quality of the roadmap that sees technologies more and more discriminately and partly takes up new topics, such as technologies for cordless communication in the 2003 issue.

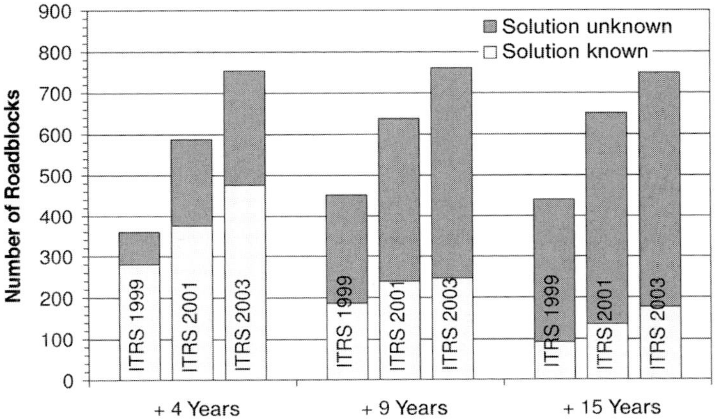

Figure 3.8 Number of roadblocks with known or unknown solutions. The figures are stated for the periods of each 4, 9 or 15 years after the publication of the respective ITRS issue (ITRS, 1999; ITRS, 2001; ITRS, 2003).

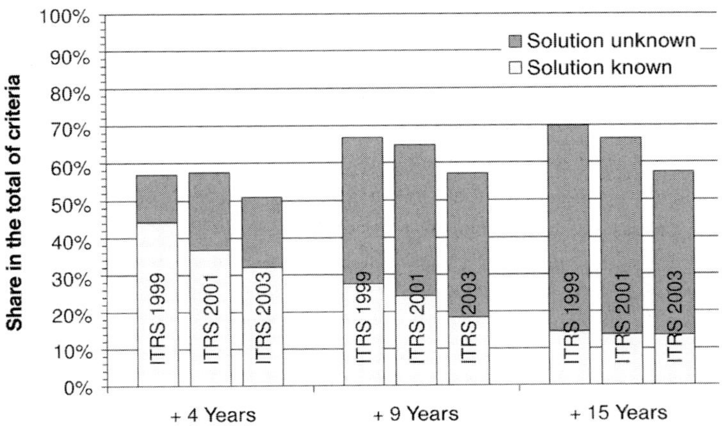

Figure 3.9 Relative proportion of roadblocks with known or unknown solutions in the total number of the criteria considered in the roadmap. The figures are stated for the period of 4, 9 or 15 years after the publication of the respective ITRS issue. (ITRS, 1999; ITRS, 2001; ITRS, 2003).

Figure 3.9 also makes clear that, seen on a long term basis (+15 years), the proportion of roadblocks with unknown solutions was even going down during the last years.

High demand for research

Thus the ITRS does not predict the end of CMOS technology. Nevertheless, it describes the immense demand for research. According to information from Infineon, Philips and ST Microelectronics, the increase of their expenditure for research and development between 1987 and 1999 was 2.5 times higher than in industry on average.

The challenges of the years to come are varied. So for insulation layers of the gate electrodes and as dielectric of the DRAM condensers, for example, materials with a high dielectric constant are required. For the insulation of the wiring of the individual components, however, materials with a low dielectric constant have to be used. In order to eliminate parasitic capacities on transistors, the Silicon on Insulator (SOI) technique was developed. The market launch of such wafers is expected in the year 2006. So-called strained silicon will offer less resistance to the charge carriers. As of about 2010, transistors with vertical structures, so-called FinFET, are probably required for the reduction of short-channel effects.

Moreover, conventional photolithography for the manufacture of future chip generations will come up against limiting factors. Research regarding new technologies, such as the EUV-Lithography (EUVL), is carried out with enormous expenditures worldwide. However, since the EUVL does not inevitably cover the demand of chip manufacturers that produce a lot of variations and small quantities per type of chip (as is often the case with ASICs, in power and communication electronics) an accompanying examination of alternative structuring techniques is carried out. In this connection, even non-optical lithographic methods or the potential of new replication methods are analyzed. Moreover, the technical utilization of self-organization phenomena is being discussed for the more distant future

The more components a chip contains, the more complex its circuit structure becomes. Therefore new design methods are being developed to realize the technological potential of CMOS technology in the design as efficiently as possible.

These are only a few roadblocks the semiconductor industry is faced with. The semiconductor industry is concerned to maintain the proven process technology in all developments. This places high demands on new materials that, for example, have to hold out under process temperatures.

Moore's 2nd Law

Apart from the technological challenges, even Moore's 2ndLaw, according to which the costs for a chip factory increase exponentially, could cast doubt on the further development along the roadmap. Figure 3.10 shows the development of these costs for the last two decades.

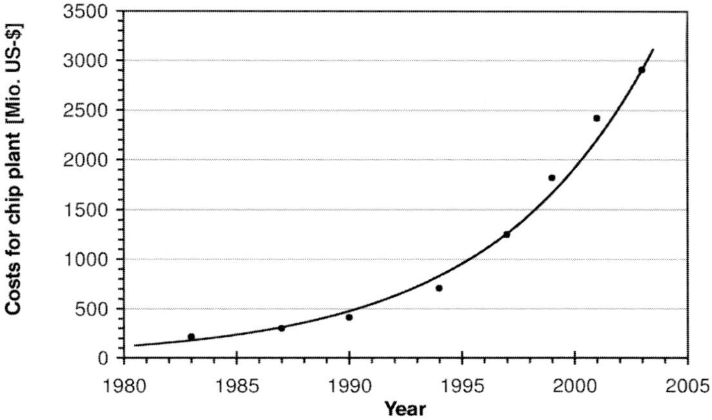

Figure 3.10 Development of the costs for a chip factory (source: Gartner Dataquest, 2003).

3.3.2
Magnetoelectronics

Magnetoelectronics is another field of nanoelectronics that has already led to successful commercialization of mass market products. Magnetoelectronics is based on the utilization of magneto-resistance effects discovered in experiments with magnetic and non-magnetic materials artificially arranged in layers. It was noticed that in such nanolayer stacks the dependence of the charge transport depends on the spin direction of the electrons. For example, with a suitable thickness of the non-magnetic intermediate layer (e.g. eight atomic layers for chrome) the magnetizations of adjoining iron layers adjust themselves spontaneously anti-parallel through quantum-mechanical "exchange interaction" over the intermediate layers. The GMR effect is based on the fact that the electron current running through the layer system is exposed to huge changes in electrical resistance depending on the alignment of the magnetic layers. Due to the layer thicknesses and the related effect alone, this effect is unambiguously attributable to nanotechnology. Within only ten years, the GMR effect discovered in Europe was used in virtually every reading head of each hard disk sold. The industrial pioneer regarding the commercial application of these nanotechnological effects was IBM.

3.3.3
Alternative Approaches

The numerous unsolved problems of silicon-based nanoelectronics will probably open up the chance to technologies less established yet, at least to occupy submarkets.[3] Figure 3.11 gives a (rough) overview of the stage of development of alternative technologies. The alternatives to CMOS examined are fundamentally different in their strategy. The SET or RTD, for example, are alternative component concepts theoretically also realizable in silicon. Some approaches, especially spintronics, would introduce new functionalities by utilizing physical effects. Other technologies, such as polymer electronics try to transfer the concepts proven for silicon to new material systems. The quantum computer, neural network and DNA computing rely on completely new kinds of information processing. Not all of these alternatives depend on the application of nanotechnology, this goes for polymer electronics, for example.

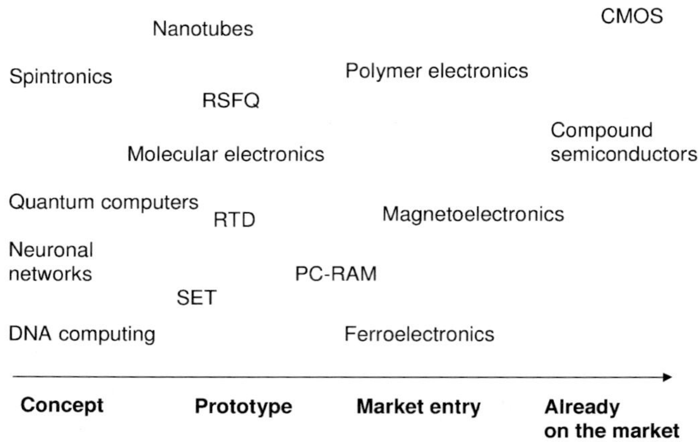

Figure 3.11 Schematic depiction of the stage of development of alternative technologies compared with the established technologies CMOS and compound semiconductors (incl. SiGe).

The hopes placed on the commercial availability of nanoelectronics are depicted in Fig. 3.12. Here such electronic components and products are concerned that are not based on a further miniaturization of the standard silicon technology, but on differing technologies with atomically exact, defined electronic functional elements on a nanometer scale. Examples for nanoelectronics are the carbon nanotube transistors or molecular electronics. Here most of the experts questioned within the context of a current commercial survey (Fecht et al., 2003) were of the opinion that nanoelectronics will gain acceptance on the market in about ten years.

3) Within the framework of this survey, it is not possible to deal with technological details of the alternative approaches. You will find detailed information e.g. in ITRS 2003; Hoffknecht, 2003 and above all in Waser, 2003.

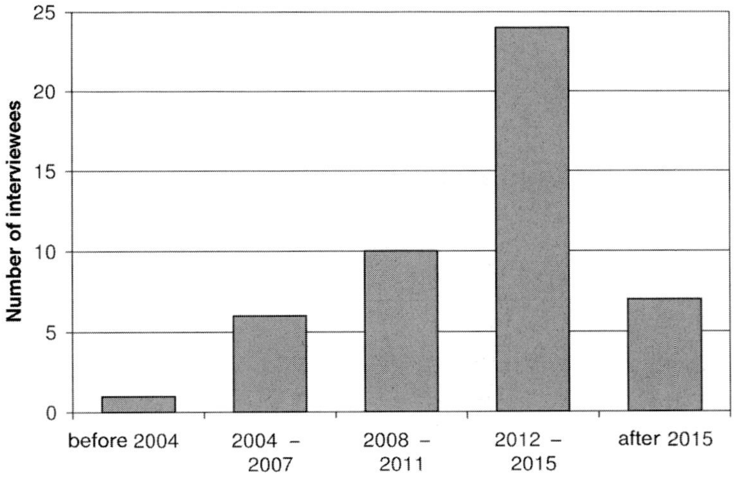

Figure 3.12 Expected time for the commercial availability of nanoelectronics based on new materials or technologies (source: Fecht et al., 2003)

3.3.4 Market Prospects

In 2002, silicon-based CMOS electronics had a market share of 98.6% according to IC Insights. The remaining 1.4% went to the compound semiconductor incl. SiGe. A mini Delphi study on the future of electronics carried out in 2002/2003 revealed that in the foreseeable future CMOS electronics will dominate semiconductor electronics. (Hoffknecht, 2003).

In the period viewed here, at best the alternative memory concepts MRAM in magnetoelectronics, FRAM in ferroelectronics and the PC-RAM will conquer first shares of the world market for chips, which is assessed at about 30 billion USD in the year 2006 (Small Times, 2003). On the other hand, these concepts will be largely integrated into the existing CMOS environment. However, for a while magnetoelectronics has already had considerable influence on the world market of hard disk drives that use GMR sensors in the reading heads. The world market for hard disk drives is assessed at 26.9 billion USD in the year 2006. Although the added value proportion of the nanotechnological component, the GMR sensor, is not quantifiable, the great leverage effect of nanotechnology on the volume market of hard disk drives becomes clear.

On a medium to long-term basis, other nanoelectronic approaches will probably be an alternative to CMOS. However, this will depend both on their own progress and on the progress of CMOS technology.

The most probable scenario will be a partial integration of new materials and components into CMOS technology, as it is already the case with FRAM and MRAM. Possible further steps on this way could be the use of carbon nanotubes as "vias" or the use of molecular electronics in so-called hybrid components (Hoffknecht, 2002).

Moreover, the established market for compound semiconductors in the field of high-frequency electronics, power electronics and optoelectronics will continue to grow, too (Hoffknecht, 2003). Nevertheless, the trend is clear that in these fields of application classically dominated by compound semiconductors, CMOS electronics is increasingly used. According to the World Semiconductor Trade Statistics (WSTS), the total market for semiconductor electronics will grow from currently 192 billion USD to 215 billion USD by 2006. The proportion of nanoelectronics will probably grow continuously to approx. 10%. Details regarding this growth will be explained in full in Chapter 6.

3.4 Nanooptics

In this survey, the field of technology at the interface between nanotechnology and optics or optoelectronics is subsumed under the term nanooptics.

According to the definition presented in Chapter 2, a relation to nanotechnology is regarded as given when function-critical key elements or structures of optical or optoelectronic elements have dimensions of a few hundred nanometers at the most or form precisions better than a few ten nanometers, or when technology is aimed substantially at the control of processes on the nanoscale.

EUV-lithography is a good example of this. The manufacture of the multilayer reflectors used there is based on the creation of thin layers with thicknesses in the nanometer range, and on the ultraprecise surface treatment. These results of nanotechnology enable the controlled handling of EUV-radiation. The control over EUV-radiation thus achieved can now be used for the generation of structures of nanoelectronics, for the quality control and measurement of which X-rays are particularly suitable in turn.

This general connection is also found in further examples described briefly in this section of the survey: ultraprecise optical components, ultrathin optical layers, nanostructured optical materials and optical systems with electric resolution in the nanometer range for inspection and structuring.

3.4.1
Ultraprecision Optics

The term nanooptics is used, inter alia, for the ultraprecision processing of optical components. Reproducible and economical production of optical components with accuracies of up to and below one nanometer is a subject of ongoing R&D. The realization of this target requires the research of new processes both for the manufacture and the measurement of such components, especially under conditions of a future economical production. In this connection it should be remembered that the term nanotechnology goes back to the ability to work on optical components with nanometer accuracies, i.e. here we are virtually faced with the birthplace of nanotechnology.

Ultraprecision optics for the manufacture of electronic components

Such precision optics is mainly used in lithography where it is urgently required for the manufacture of electronic components of increasingly smaller structural sizes. In lithography, the quality of optics is given top priority and leads to the limits of the currently technically feasible, resulting in correspondingly high manufacturing costs and thus in a high price. The world market volume for such lithography optics is assessed at 0.5–1 billion EUR for the year 2006 (source: own company survey). Further technical challenges in this connection are the nanometer-precise assembly, adjustment or stabilization of the whole system as well as the system design in consideration of even the smallest deviations on the nanoscale.

Moreover, increasingly smaller wave lengths require the application of mirror optics instead of transmitting lens systems. In addition, extremely complex and nanometer-perfect layer systems have to be applied in order to achieve the necessary functionalities.

Far less drastic are the requirements for products in the consumer product range, like aspheres for data projectors, cameras, lenses, scanners etc. Here, above all the need is for efficient manufacture and measurement of optics at moderate precision.

Current problems of nanooptics are aimed at the mastering of geometries with low symmetry, subsumed under the term aspheres. This starts at weak deviations from the spherical shape of only a few percent and leads over cylindrical forms up to free-form surfaces of almost any geometry. Therefore, the field of ultraprecision processing will have a significant share in interdisciplinary approaches in the sector of optical technologies even in future. However, it is expected that the relevance of genuine nanotechnological problems will increase rather than diminish.

3.4.2
Ultrathin Optical Layers

A subset of nanooptics may also be ultrathin optical layers and layer systems as well as methods for the application of such layers. Such layers have a central function in reflection optics for EUV lithography, for instance. Further applications are found in optoelectronic components as transparent protective layers and antireflex layers.

3.4.3
Measurement Technology

Optical sensors for measurement and analysis systems with nanoscale resolution

With the growing importance of ultraprecise components, innovative measurement methods are also required for the nanometer-perfect characterization and certification of such components, and are classified as nanooptics. Optical sensors and sensor systems based on nanotechnological function principles as parts of measurement or analysis systems are possible new products, just like devices that supply information via electromagnetic fields in visual (and near IR and UV) frequency ranges, with a nanometer-perfect lateral resolution. The world market volume for optical sensors and sensor systems is assessed at 1–5 billion EUR for the year 2006 (source: Workshop Optics).

Fields of application are surface characterization, positioning and localization, characterization of interactions between light and matter, the characterization of nanostructures and their functional properties as well as the verification of tolerances in the nanometer range.

3.4.4
Microscopy

Microscopic processes with respectively high resolution are a particularly important group of the measurement processes mentioned before. With new microscopy methods, such as confocal microscopy, near-field microscopy or UV-microscopy, it is possible to depict and manipulate nanostructures.

Near-field microscopy is based on near-field optics, the essential feature of which is its ability to overcome the diffraction limit and the fact that with this, resolutions below the utilized light wave lengths are to be achieved.

Especially in the field of biophotonics, i.e. in the micro-characterization with utmost resolution of structural, functional, mechanical, biological and chemical properties of biological materials on a subcellular level, certain microscopic processes belong to nanooptics. So with the 4-pi-confocal microscopy in combination with the Stimulated-Emission-Depletion (STED) and the Point-Spread-Function Engineering, there are good prospects of reaching resolutions below 100 nm. Apart from this, methods for single-molecule spectroscopy and the fluorescence-energy transfer spectroscopy, which are also assigned to nanooptics, play an important role.

3.4.5 Photonic Crystals

Nanooptics includes completely new optics concepts, such as photonic crystals, in which the realization of a so-called band gap for light is feasible by suitable nanostructuring. With this, it is possible to lead and manipulate the light in a very confined space. Photonic crystals are materials in which the dispersion of electromagnetic radiation is specifically and massively influenced by the periodic variation of the refractive index. In particular, there are frequency ranges in which wave propagation is impossible. In analogy to the semiconductor physics, they are called band gaps. The properties of photonic crystals open up completely new prospects for the manufacture of compact optoelectronic components and for optoelectronic integration. Applications are based essentially on few basic elements that are achieved by the integration of localized defects into the photonic crystal. These basic elements are waveguides, kinks, resonators and extremely dispersing prisms. They are considered for different applications in the optical communications technology.

Photonic crystals with a certain band gap require sufficiently regular and defect-free structuring of a material in the range of wavelengths of the guided radiation as well as an efficient fiber coupling system. Since these problems have not yet been solved satisfactorily for any material class, both classical semiconductor materials and high-refractive glasses and polymers are being examined for their suitability as components of photonic crystals.

Photonic crystals open up new prospects for optoelectronic components

Further applications of photonic crystals may be found, inter alia, in the lighting and gas sensor technology. An area of investigation is the linking of the active area of a light-emitting diode (LED) with a spot defect in a photonic crystal in a way that the LED can emit only in one single resonator mode. This would enable the production of LEDs with very high light coupling efficiency. For applications in gas sensor technology, photonic crystals could be used to increase the time that infrared light needs to pass through the absorption cell, which would make the realization of more compact gas sensors possible.

3.4.6
Optoelectronic Light Sources – Lasers and Light-emitting Diodes (LED)

Lasers and LEDs of nanometer-thick semiconductor layers drive innovation in many sectors of industry

Another important aspect of nanooptics is to be seen in the exploration of new types of semiconductor light sources, i.e. lasers and LEDs. These optoelectronic components generate light in extremely thin, only nanometer-thick semiconductor layers. They represent one of the few examples of a bottom-up approach, rather than the miniaturization of a known technology, that led to the launch of new types of products connected with enormous economic success. Further developments in the years to come are directed towards the development of new wavelength ranges and the enhancement of light output, efficiency and working life.

Applications of such light sources are, for example, displays, optical data storage or lighting, but also detectors and passive components. The world market volume for optoelectronic light sources in the year 2006 is assessed at 1–5 billion EUR for diode lasers and at 1–5 billion EUR for LEDs. Here, the proportion of "white" LED is estimated at less than 50 million EUR per year (source: Workshop Optics).

3.4.7
Quantum Dot Laser

A less developed example for a new optoelectronic light source is the quantum dot laser.

In the epitaxy of a few monolayers of a semiconductor on a semiconductor substrate with different lattice constants (heteroepitaxy), isolated, arranged islands develop depending on the growth conditions. Each of these islands displays a relatively accurately defined form and size in the range of 10 nm; they show specific electric and optical properties, such as discrete energy levels, so that they are often called quantum dots. Lasers can use quantum dot layers or stacks of quantum dot layers as an active medium. Theoretically, quantum dot lasers show the potential to exceed the efficiency of conventional quantum film lasers with regard to certain operating parameters.

3.5
Nanobiotechnology

Nanobiotechnology as an interface between nanotechnology and biology came into the focus of attention recently because of pioneering scientific work on the functioning principle of motor proteins, the structure of electronic components using DNA and the application of nanoparticles in the treatment of illnesses. Within nanotechnology, nanobiotechnology is regarded as one of the most promising fields in future, with high economic potentials especially in the health sector (Georgescu and Vollborn, 2002). Because a wide range of application options and research problems is being worked on in nanobiotechnology, they are divided into two fields according to the application potential:

Two branches of research: "Bio2Nano" and "Nano2Bio"

- Bio2Nano
  Here, technologies are subsumed that aim at the manufacture of nanotechnological products utilizing biomolecules.
- Nano2Bio
  The field of Nano2Bio comprises applications of nanobiotechnology in life sciences, in particular medical and pharmaceutical applications.

3.5.1 **Bio2Nano**

The current research approaches in the application field of Bio2Nano can be summarized in the following categories (comp. Wevers and Wechsler, 2002):

- Nanostructuring through self-organization processes (e.g. self-organization of DNS-molecules and bacterial membrane proteins or nanostructuring through biomolecules as templates)
- Material synthesis through biomineralization (manufacture of biological, functional materials through self-organization processes)
- Biosensors and biomembranes (e.g. in environmental technology, production and food engineering)
- Biomimetic energy generation (e.g. biologically supported photovoltaic, biomimetic fuel cell)
- Biomolecular motors and actuators (e.g. the microtubuli-kinesin delivery system)
- Application of biomolecules in technical products (e.g. data storage, molecular electronics, fake protection, DNA-computing, neurotechnology)

In the field of utilizing biological materials and processes, the topical complex of "technical utilization of self-organizing phenomena" will be particularly important in future, inter alia as a possible alternative to conventional lithographic methods (for this, particularly long research lead times are required). The same goes for development and use of nanodimensional machine technologies that are considered to be of importance. This includes the wide range of cell-free motion models (e.g. protein motors) which may be used for nanoscale manipulation, controlled movements of objects or specific substance delivery. The possible applications are found mainly in biotechnological, biomedical and chemical fields.

Nanobiotechnology will play an important role for the field of development and utilization of biological and biomimetic materials with completely new properties. Here, technically conditioned structuring methods combine with processes of self-structuring and self-organization using biological principles. In the case of structuring, among other things the development of economic alternatives to conventional lithographic methods is possible. Further examples of predicted product are metalized layer proteins – as template for highly active catalysts or extremely sensitive gas sensors – or photochrome proteins (e.g. bacteriorhodopsin) as a universally and economically applicable copy protection.

Research approaches of nanobiotechnology mostly still in the state of basic research

However, apart from a few exceptions, the research approaches described are still at the stage of basic research. In some fields, conclusive proof of the operational capability of the underlying principle has been provided (proof-of-principle). The essential preconditions for conversion into marketable products, however, do not yet exist in most of the current research approaches. These requirements are reproducibility and controllability of the production process; stability and durability of the biomolecules and systems used; specific application relevance; and better performance in comparison with alternative products and processes. Therefore, in most cases a long-term time horizon (ten years and more) is envisaged for possible conversion into marketable products, except for a few cases, such as the use of bacteriorhodopsin as copy protection and simple data storage systems or the application of S-layer-based sensor systems and catalysts. It is not yet possible to make a serious assessment of the market potential in the application field of Bio2Nano, because the developments are still too fundamental and it is still unclear whether the new methods and techniques can stand up to the existing competition of proven fields of technology.

3.5.2 Nano2Bio

Nanotechnological applications for medicine and pharmacy include recent intensive research activities in the fields of drug delivery and new kinds of biochip systems (Wagner and Wechsler, 2004). Furthermore, the application of nanoparticles in molecular diagnostics and the use of nanostructured materials for the creation of bioactive surfaces are active areas. In tissue engineering and in neuroprosthetics, there are currently only individual examples for the use of nanotechnology, and its future importance in these fields still has to crystallize.

Nanotechnological concepts are currently employed in the control of biological processes, for instance, in the application of functionalized nanoparticles as a new kind of technique for local drug delivery, for controlled nanostructured prostheses and implants with improved biotolerance and for innovative processes for molecular and cellular diagnostics by means of nanostructured surfaces and ultrathin layers. Current issues of nanobiotechnology are particularly directed at the mastering of biological–technical interfaces, so-called "interface engineering" or interface design. The controlled handling of cells and cell associations requires suitably nanostructured and functionalized surfaces and membranes. Besides the field of "tissue engineering", particularly pharmacology will profit from this. It is foreseeable that interface design will be an important element of innovative techniques for in vivo validation of drug targets. Here, the intended aim is the provision of better processes for the faster and more specific testing or validation of medical drugs.

Functionalized nanoparticles for applications in pharmacy and medical technology

Another medium-term priority is the field of "functional biohybrid systems". This concept expands the focal subject of "interface engineering" to the point where cells or tissues are not only suitably stored, treated or characterized, but also actively functionalized. Particular fields of application are neuroactive implants, the exploration and/or treatment of neurodegenerative diseases as well as neurotechnology. In this respect, the connection of electronic and biological systems has a key position. Success in this field is an essential precondition for pushing the door open to neuroelectronics. Here, the mastering of nanotechnological structuring and manipulation techniques can make a decisive contribution. With this, it would be possible to reach the physical limits of miniaturization of electronic components or systems.

3.5.2.1 **Basic Biomedical Research**

One of the reasons for the rapidly growing importance of nanobiotechnology in medicine and pharmacy is the continuous further development of instrumental analysis that allows the examination of biological objects on the nanoscale. Special scanning probe techniques and the optical single molecule detection are examples that enable the examination of single cell components in their natural environment, thus contributing much to the understanding of the functionality of the cell. In particular, information about the function of the proteins and the signal path within the cell may lead to new strategies for the combat of diseases.

3.5.2.2 Drug Delivery

Nanoscale drug delivery systems offer the potential:

- to deliver poorly soluble or chemically unstable drugs in aqueous media to damaged tissue;
- to overcome biological barriers, such as the blood–brain barrier;
- to accumulate drugs specifically in the damaged tissue to diminish the danger of side effects.

First medicines with nanoscale drug delivery system on the market

After more than 20 years of intensive research in this field, the first medicines making use of such drug delivery systems are on the market. This is regarded as a field of application for nanobiotechnology with a large economic potential, as there is a multitude of candidate drugs that can only be administered by a suitable nanoscale delivery system.

3.5.2.3 Contrast Media in Diagnostics

Nanoparticles can be used for molecular imaging in order to accumulate contrast media in damaged tissue using molecular markers. Since the molecular signatures of many diseases appear even before the symptoms break out, these methods enable the diagnosis of diseases in the early stages. This is accompanied by a paradigm shift in medicine, such that its focus is going to shift increasingly from restoration to preservation of health.

3.5.2.4 Biochips

Another field in which nanotechnology is gaining importance is biochip technology. The trend towards the miniaturization of DNA-chips is still unbroken, so that the lateral resolution continuously approaches the nanoscale range. For research applications, nano-arrays have already been launched. Apart from the high-density DNA chips mainly used for expression studies and SNP analyses, there is also a market segment for biochips with lower spot densities, which are used in biomedical research for the examination of certain symptoms and which will be used in medical diagnostics in the future. Even for protein analysis, it is the protein chips with a relatively low spot number that are more likely to gain importance. In this field, nanotechnology will not only play a role in the further miniaturization, but also in the improvement of detection sensitivity and reliability of the systems. For example, nanoparticle fluorescence marker or near-field optical detection systems are used to increase the detection sensitivity of biochips.

Apart from this, a lot of new types of electric and magnetic biochip detection systems based on nanotechnology are currently being developed. German research institutes and companies are very active in this field and internationally well positioned. In comparison with conventional optical processes, electric and magnetic detectors are more robust and easier to integrate into a miniaturized sensor. On a medium or long-term basis, such compact systems will open up a mass market for medical diagnostics, in particular to replace expensive and time-consuming laboratory tests by quick tests on the spot at the medical practitioner or in hospital. Such systems are also being developed for application in personalized medicine. Here the aim is, e.g. to design medication specifically to the patients' needs and to optimize therapy.

Mass market in medical diagnostics possible

3.5.2.5 **Implants**

Another application area of nanobiotechnology is the surface structuring of implants to improve their ingrowth. Great importance is attached to the better understanding of the processes at the interface between the tissue and the surface of the implant. Here, it is essential to understand how cells react on nanoscale structures in their environment. These topics are also very important for tissue engineering, as surfaces of tissue matrices have a decisive function in the control of the cell growth that is not yet well understood.

Also classed with nanobiotechnology are the "intelligent implants", such as neuroprostheses or implantable drug delivery microchips. From these implants, however, only individual components can really be assigned to nanotechnology. Nanostructured surfaces are used to improve the biocompatibility of the implant or the ingrowth of electrodes. As far as their lateral dimensions and their functional components are concerned, intelligent implants come less within the remit of nanotechnology, but more of microtechnology. This reveals a frequently seen principle of nanotechnology: that although in many products it has only a very small share in the added value, certain product features can only be realized using nanotechnology. This causes a leverage effect, as the small share of nanotechnology leads to a high added value of the product and thus generates the respective competitive advantage.

Nanotechnology has a strong leverage effect on the added value of medical products

Only a few market analyses about applications of nanobiotechnology in medicine and pharmacy are readily available. According to a study by the Business Communications Company (BCC, 2003), the global market volume of products attributable to nanobiotechnology is stated to be 269 million USD in the year 2002; for the year 2007, turnovers are assessed at 1.2 billion USD. According to the study, the highest turnover, almost 200 million USD in the year 2002, was realized with instruments of biophysical analysis, such as scanning probe techniques.[4] The market volume of nanotechnological products for

4) The global turnover for scanning probe microscopes is given without considering the fact that these instruments are also used for non-biological applications.

Table 3.7 Market potential of nanobiotechnology in medicine and pharmaceuticals according to a study by the Business Communication Company (BCC, 2003).

	Turnover worldwide/ million USD		Number of companies taken into account, 2002
	2002	**2007**	
Biophysical analysis (e.g. scanning probe techniques)	181	745	27
Diagnostics and analysis (e.g. nanoparticles for biochips)	80	391	57
Active substances and drug delivery	8	33	33
Tissue engineering	0	1.5	7
Total	**269**	**1171**	**124**

medical analysis and diagnostics, which also includes contrast media or nanoparticle markers for biochips, is stated at 80 million USD. According to this study, tissue engineering products, produced with a contribution from nanobiotechnology, were still not on the market in the year 2002.

When interpreting market surveys, it must be taken into account that the market analyses these surveys are based on do not analyze the whole market exhaustively. Moreover, they are based on different definitions of nanobiotechnology, so that whole product groups are left out of consideration, such as active substance nanocrystals, because their degree of innovation is considered insufficient , or certain liposomal active substances, because they are larger than 100 nm. This is the reason why the BCC survey states a turnover of only 8 million USD for the drug delivery sector in 2002, although there are already a lot of products on the market that today reach a sales volume of several 100 million USD each year.

In one of their surveys, Front Line Strategic Consulting (2003) gives a clearly more optimistic assessment of the total market for nanobiotechnological products than BCC. For the year 2008, they expect a turnover of 3 billion USD in this field, a figure that includes, apart from the life science sector, the considerably smaller market for technical applications as well.

Although such surveys cannot exactly forecast the market volume, at least they reflect an approximate idea of the economic importance of a technology. Based on the surveys now available, it can be assumed that by the end of the year, sales of nanobiotechnological products will have reached several billion US dollars.

3.6 Nanotools and Nanoanalysis

The nanotools and nanoanalysis sector comprises all the equipment applied to the manufacture and characterization of structures in the nanometer range. It can be divided into the following categories, described below:

- Devices for lateral nanostructuring (lithography)
- Devices for the manufacture of nanoscale layer systems
- Analytical equipment for the characterization of nanoscale structures
- Nanometrology and positioning

3.6.1 Creation of Lateral Nanostructures

The creation of lateral nanostructures is a compulory precondition for the manufacture of competitive products, especially in electronics. The most competitive nanostructuring method by far, seen from the economic point of view, is photolithography as the basis for the manufacture of CMOS-based electronic components. Alternative methods, such as soft lithography, are mainly considered for special or niche applications.

3.6.1.1 Optical Lithography

In optical lithography, nanostructures are generated through light rays projected by means of structured masks onto a substrate surface that is coated with a photoresist. After the development of the resist, the projected structure is transferred by means of etching processes, for instance. Successive repetition of the procedure results in highly complex structural patterns on the silicon wafer forming the basis of electronic components.

Different approaches for lithographical processes of the next generation (NGL)

With the efficient photolithography devices currently applied, the creation of structures of 90 nm is possible, and thus they are assignable to nanotechnology by definition. For further miniaturization, various consortia, in collaboration with chip manufacturers (IBM, Intel, etc.) and manufacturers of lithography devices, are examining lithography processes for the next generation (NGL) that will allow structuring sizes of 90 to 35 nanometers in batch production. This includes, inter alia, the extreme UV and X-ray lithography, but also other procedures such as the SCALPEL technique (Scattering with Angular Limitation Projection Electron Beam Lithograhy, an electron beam method further developed by Lucent Technologies), PREVAIL technique (Projection Reduction Exposure With Variable Axis Immersion Lenses, a version supported by IBM and Mitsubishi) and the ion projection lithography of Infineon (Service 2001). The mainstream of nanolithographic developments is currently focused on EUV lithography, while alternative processes, such as switchable masks and maskless processes are considered more in connection with special applications with small numbers of pieces (e.g. ASIC).

The market value of the equipment required for photolithography is in the order of several billion USD per year. Therefore, the world market for lithography steppers (tools) is assessed at 7.7 billion USD and for mask making devices at 0.9 billion USD for the year 2006 (electron beam lithography) (Fecht et al., 2003).

3.6.1.2 **Soft Lithography**

Soft lithography as economical nanostructuring method

Optical lithography equipment demands immense investment, so soft lithography methods are being developed as alternative and more economical nanostructuring methods, particularly for special applications with small numbers of pieces. Here, the term nanoimprint lithography subsumes different printing or embossing techniques (hot embossing or microcontact printing), which can help to produce micro and nanostructures in different material systems, such as organic polymers or inorganic solids. After the nonrecurring manufacture of a high-precision printing pattern (e.g. by means of electron beam lithography in the sub-100 nm range), nanoimprint lithography has the potential for a parallel high-throughput production process. In the case of the variant of UV-nanoimprint lithography developed by the Advanced Microelectronic Center Aachen (AMICA), a nanostructured printing pattern of quartz glass is written by means of optical or electron beam lithography, depending on its size, and then transferred to the printing material by means of conventional etching processes. The printing process occurs by pressing the stamp into a thin polymer (resist) spun onto the substrate. Then it is hardened through the transparent stamp by means of UV light. Afterwards the stamp is removed and the resulting polymer relief is used as component or as etching mask for the following structure transfer into the substrate. The Swedish Obducat is regarded as the market leader in the field of nanoimprint technology – a topic German companies are dealing with, too. The German enterprise Mildendo, for example, concentrates on a hot stamping method that allows the forming of nanostructures up to 200 nm (mainly for channels in microfluid systems) with extremely high structural accuracy.

Soft lithography also comprises serial methods such as the so-called dip-pen lithography. Here the application of nanoscale structures occurs directly onto a substrate. The point of an atomic force microscope wetted with writing liquid is used to "write" on a substrate surface. In the wetted area, the molecules applied to the substrate organize themselves spontaneously in thick arrangements with widths of approx.

10–15 nm, analogous to self-organized monolayers. With the DPN technique, for instance, specific binding spots on a substrate can be developed that may be used for biotechnological applications (DNA-chip arrays etc.).

3.6.2
Development of Nanolayer Systems

Special surface coating methods enable the development of ultrathin functional layers with characteristic layer thicknesses of less than 100 nm that lead to new types of or improved product functionalities. Nanoscale layers are already used as functional elements in many fields of applications, as in the optical industry, in electronics/opto-electronics or in machine engineering (see Table 3.8). The description of the equipment used for the development of nanolayer systems should be limited to vapor phases and vacuum processes, since here, in contrast to the wet-chemical route, a rather complex and with this marketable equipment is used. The main distinction is made between chemical (CVD) and physical vapor deposition (PVD). There are so many different process types, that the depiction is limited to some selected variants.

Table 3.8 Application fields and material systems of different vapor-phase deposition methods.

Method	Fields of application	Material systems
CVD • PECVD • MOCVD • Photo-CVD	**I&C-Technology** • compound semiconductors for LED • laser, signal amplifiers • transistors in the field of telecommunication/optoelectronics **Automotive engineering** • sensors (GPS, radar), LED **Energy engineering** • compound semiconductors for solar cells	• II-VI-, III-V- materials • oxides, metals and dielectrics • ferroelectrics • HTSC • optical coatings (for mirrors, filters)
PVD • DC magnetron sputtering • MBE • Ion implantation	**Machine Engineering** • hard coatings • friction reducing layers • decorative layers • corrosion protection • wear protection • tool coating **Electronics, optics** • diverse components in the fields of optoelectronics and magnetoelectronics, ultraprecision optics	• II-VI-, III-V- materials • oxides, metals and dielectrics • ferroelectrics • HTSC • hard coatings • optical coatings

3.6.2.1 PVD Process

A great number of methods are covered by the field of PVD (Physical Vapor Deposition) processes, mostly used in high vacuum, such as vacuum deposition (layer formation, inter alia by Molecular Beam Epitaxy etc.), sputtering (removal of atoms or atomic clusters from the surface of a solid by ion bombardment) and ion plating (specific modification of properties in thin surface layers by the implantation of ions). These processes offer a wide range of possibilities for the modification of deposition parameters and allow the coating of most different materials such as steels, metals, semiconductors, plastics, ceramics or glass. The world market for PVD equipment is assessed at 2.1 billion USD in the year 2006 for the sputtering process, at 1.4 billion USD (Fecht et al., 2003) or 3.1 billion USD (BCC, 2002) for ion implantation and at 1.1 billion USD for the MBE process (BCC, 2002). According to BCC, annual growth rates in this sector will range between 25 and 33% by the year 2006. German companies working in the field of manufacturing of PVD equipment for nanocoatings are VTD Vakuumtechnik, Leybold Optics or Balzers Verschleißschutz GmbH, for example.

3.6.2.2 CVD Process

In contrast to the PVD methods, thin layers are produced during the CVD processes, by a chemical reaction of gaseous source materials ("precursers") brought into the reactor in an inert carrier gas. CVD processes are used, for example, for the development of Si-Wafers, compound semiconductors for lasers and LED or even for the production of carbon nanotubes. The world market for CVD equipment is assessed at 5.7 billion USD for 2006, with annual growth rates of 12% (Fecht et al., 2003).

3.6.3 Nanoanalysis

Nanoanalysis comprises a multitude of highly-developed, partly long-established procedures suitable for characterization of objects on the nanoscale. For measurement purposes, diverse physical, chemical and biological interactions and effects are used. A central role in nanoanalysis is attributed to scanning probe techniques (SXM methods). These are microscopy techniques with atomic resolution in which the substrate is raster scanned by means of a microprobe tip, and the physical and chemical interactions arising between substrate and probe are measured. There is a large number of different variants of scanning probe techniques, such as the Atomic Force Microscopy (AFM) the Scanning Tunnel Microscopy (STM) or the Scanning Near Field Optical Microscopy (SNOM). Important fields of application of the scanning probe techniques are materials research, the manufacture of semiconductors and the life science sector. Scanning probe techniques are used for, for instance, quality controls in production processes, inter alia, of extremely thin layers. NanoSensors, a company founded in Germany but now resident in Switzerland, has specialized in silicon cantilever tips for scanning probe microscopy and is regarded as the leading manufacturer worldwide. World market leader among the manufacturers of equipment for 3D metrology and production control that allows scanning of the surface structure up to the nanoscale (nanometrology) is Veeco, a company which, among others, bought Digital Instruments (another company leading in scanning probe methods under ambient conditions) and Thermo Microscopes in 1998. The German company Omicron Nanotechnology has specialized in surface analysis by means of scanning probe microscopy in ultrahigh vacuum and meanwhile has become the leading supplier of appliances and equipment in this sector worldwide. The development of the world market for scanning probe equipment is forecast to increase from currently approx. 200 million USD to 800 million USD in the year 2007 (Small Times, 2002).

Another important method for the examination of nanoscale or atomic structures is electron microscopy with its main fields of application in materials research and life sciences. However, this method involves high demands on the test preparation and great technical effort. The world market for SEM equipment is assessed at 0.6 billion USD for 2006 (Fecht et al., 2003).

Other methods for the characterization of nanostructured surfaces and films include, inter alia, elipsometry or the nanoindentation procedure that allows the analysis of mechanical features of thin films and coatings without expensive test preparations through the precise, nanometer-deep indentation of a diamond point and the ensuing measurment of changes. Important parameters to be determined are, for instance, texture, morphology or defects in nanolayers. Application fields are the quality control in microsystem engineering and semiconductor manufacture, in the health sector for the determination of the mechanical integrity of metal implants, and in the consumer goods sector, e.g. for quality controls of razor blades and cosmetics. The world market for thin-film measuring techniques is assessed at 0.5 billion USD for 2006 (Fecht et al., 2003).

Furthermore, there are a great number of methods applied to the characterization of nanomaterials, the main application field of which, however, lies more beyond nanotechnology, such as molecular spectroscopic methods or X-ray analysis.

3.6.4
Ultraprecise Surface Processing

Ultraprecision technology comprises all processing methods for the extremely precise development, in form and smoothness, of bodies and surfaces with macroscopic dimensions. The more precisely the surfaces are smoothed out and formed, the better are the functional (e.g. optical) properties that can be achieved. The most important procedures of ultraprecise forming or form correction include the mechanical/chemical and optical processing methods as well as ion beam and plasma processing methods. Ion beam and plasma processing methods allow form correction or forming on large surfaces (cm^2 to m^2) with a depth precision in the nanometer range and the reduction of roughness to sub-nanometer values. The low processing speed and the high plant costs, mean that the application range in industrial manufacture is limited to high-efficiency optics. Optical methods, especially with the use of UV lasers, are applied to the ultraprecise processing, e.g. of polymer surfaces. Application fields of ultraprecise surface processing are found, inter alia, in optics. Here, apart from lenses that have to be ever smoother and more precise in form for the visible sector, optics for the infrared and even the UV and X-ray sector are increasingly required.

Ultraprecise surface processing also plays an important role in bonding technology in microelectronics. For the economic manufacture of microsensors and actuators the so-called direct bonding of silicon wafers and other components is gaining more importance. This concerns both the bonding of silicon and other semiconductor elements (optical elements based on III/V semiconductors) on a chip and the assembly of different optical and mechanical microcomponents (e.g. microlenses of quartz glass, piezoelectric actuators etc.). With direct bonding, two ultraprecise plain surfaces are brought into contact so that they can be bonded irreversibly by one pressure and temperature step without additional adhesives. The world market for plants for ultraprecise surface procession is assessed at 250–500 million EUR in 2006 (source: company survey).

3.6.5
Nanometrology and Positioning

Nanometer-precise positioning and control are compulsory conditions for production processes, especially in microelectronics and system engineering. Different procedures have been developed by different companies such as Piezomax Technologies Inc. (Vancouver), Klocke Nanotechnologie (Aachen), Kleindieck Nanotechnik (Reutlingen) or Physik Instrumente GmbH (Waldbronn). Nanowave has developed a patented method of ultraprecise positioning (sub-nanometer range) on the basis of scanning probe microscopy in combination with high-frequency probe oscillation. The position of the probes is determined by the comparison with a reference frequency. This system is applied in lithography steppers, for example. An important application field of nanoprecise manufacture is the production of semiconductor components for electronic data storage and optical data transfer. The world market for products in the field of nanopositioning is assessed at 0.5 to 1 billion EUR for the year 2006 and at 10–50 million EUR in the field of nanorobotics (source: company survey).

3.7 Overview of Market and Applications Potentials

Table 3.9 summarizes the market potentials of nanotechnological applications of the different partial disciplines described above. It is not possible to evaluate the "world market for nanotechnology" on the basis of the figures given, since

- market data is only available for a part of the nanoproducts, hence the lists are incomplete;
- market forecasts partly refer to different time horizons;
- nanotechnological products are named twice in two or more sections (e.g. application of nano-basic products/components in products from other sectors).

Table 3.9 Assessment of the annual world market volume of nanotechnological products.

Nanotechnological products	Annual world market volume (reference year)
Nanomaterials	
metal oxide /metal nanoparticles	900 million USD (2005)[1]
nano-silicic acid	800 million EUR (2003)[2]
nano-layer silicates	25 million EUR (2006)[3]
CNT	145 million EUR (2005)[4]
	1.2 billion EUR (2006)[5]
carbon black	3 billion USD (2002)[6]
	8 billion USD (2006)[5]
polymer dispersions	15 billion EUR (2002)[7]
organic semiconductors	500 million USD (2005)[10]
dendrimers	5-15 million EUR (2006)[3]
micronized active substances	1 billion EUR (2002)[7]
zeolites	2.6 billion USD (2006)[5]
aerogels	10 billion USD (2005)[8]
polymer nanocomposites	0.3 billion USD (2006)[3]
	1.1 billion USD[5]
	1.5 billion EUR (2009)[9]
Nanolayers	
hard layers	0.5–1 billion. EUR (2006)[12]
tribological layers	1–5 billion EUR (2006)[13]
antifog layers	50–250 million EUR (2006)[12]
tool coatings	50–250 million EUR (2006)[12]
corrosion protection coatings	1–5 billion EUR (2006)[13]
electronics based on functional nanolayers, e.g. GMR-HDD	>5 billion EUR (2006)[13]
Nanobiotechnology	
biophysical analysis (e.g. scanning probe techniques)	181 million USD (2002) 745 million USD (2007)[14]
diagnostics and analysis (e.g. nanoparticles for biochips)	80 million USD (2002) 391 million USD (2007)[14]
active substances and drug delivery	8 million USD (2002)
	33 million USD (2007)[14]
tissue engineering	0 million USD (2002)
	1.5 million USD (2007)[14]
Nanooptics	
lithographic optics	0.5–1 billion EUR (2006)[12]
ultraprecision optics	1–5 billion EUR (2006)[13]
LED	1–5 billion EUR (2006)[13]
thereof white LED	10–50 billion EUR (2006)[13]
diode lasers	1–5 billion EUR (2006)[13]
thereof high-efficiency diode lasers	50–250 billion EUR (2006)[12]

Table 3.9 Continued

Nanotechnological products	Annual world market volume (reference year)
Nanoelectronics	
CMOS-electronics <100 nm	20 billion USD (2006)[18]
GMR-HDD	26.6 billion USD (2006)[5]
MRAM	30–50 billion USD (2010)[19] (if applicable, replacement for DRAM)
Nanotools/Nanoanalysis	
lithographic steppers	7.7 billion USD (2006)[15]
electron beam lithography	0.9 billion USD (2006)[15]
sputter processes	2.1 billion USD (2006)[15]
ion implantation	1.4 billion USD (2006)[15]
MBE procedures	1.1 billion USD (2006)[15]
CVD equipment	5.7 billion USD (2006)[16]
scanning probe microscopy	200 million USD (2002), 800 million USD (2007)[17]
SEM	0.6 billion USD (2006)[15]
thin-film measurement technology	0.5 billion USD (2006)[15]
ultraprecise surface processing	250-500 million EUR (2006)[12]
nanopositioning	0.5–1 billion EUR (2006)[12]
nanoparticle counter	10–50 million EUR (2006)[12]
nanorobotics	10–50 million EUR (2006)[12]

Sources: [1] BCC, 2002, [2] Wacker Silicones, 2003, [3] SRI, 2002, [4] Mitsubishi Research Institute, 2002, [5] Fecht et al., 2003, [6] Reuters, 2002, [7] BASF/Distler, 2002, [8] Aspen Systems, 2001, [9] Stevenson, 2002, [10] Frost&Sullivan, 2002, [11] Frost&Sullivan, 2003, [12] company survey, [13] VDI TZ-Expert-Workshop, [14] BCC, 2003, [15] VDI Nachrichten, 2003, [16] BCC, 2002b, [17] Small Times, 2002, [18] own assessment, [19] Small Times, 2003.

4
Indications of Market Potentials in Patent Data

Dirk Holtmannspötter

4.1
Methodological Preliminary Considerations

4.1.1
Patents as Indicators for Economic Potential

If patent information is to provide indicators of the economic potential of a technology, then the fundamental question regarding the value of patents must first be discussed. It is a known fact that patents can have a high value and play a central role for the competitive position of companies[1]. Far less often considered is the fact that many patents are of hardly any value. However, a fact that applies to all patents is that they run up costs. Scientific literature describes a lot of methods for the evaluation of patents in different situations (Wurzer, 2002; Pitkethly, 2002; Reitzig, 2002). Patent evaluation is necessary to support management decisions during R&D projects and for their exploitation, such as:

Patent evaluations

- Should an invention be patented?
- Should a patent application once initiated be followed up? In which countries?
- Should a patent once granted be maintained? In which countries?

However, patent evaluations are also important within the framework of license negotiations, for the sale of patents, in negotiations regarding venture capital for the start-up of companies, the sale and merger of companies, as well as for the balancing of accounts and reporting.

In the end, the value of patents is determined according to the future market value of the related products. Accordingly, the patent eval-

1) A current example is the decision of the Tokyo district court, which obliges the Nichia company to pay 180 million USD to Prof. Nakamura, the inventor of the blue LED (Normile, 2004).

Growth Market Nanotechnology. Edited by N. Malanowski, T. Heimer, W. Luther, and M. Werner

ISBN: 3-527-31457-1

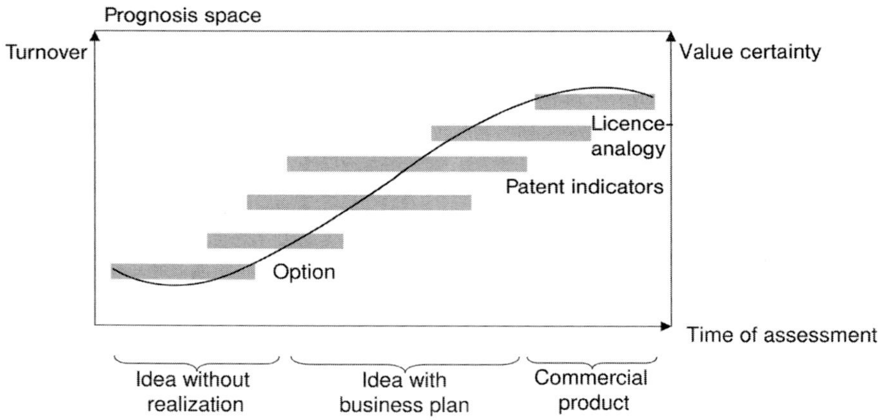

Figure 4.1 Methods for the evaluation of patents through the technology life cycle (according to Wurzer, 2002).

Predictive uncertainties

uation itself is based on some kind of market forecast and is therefore subject to all the uncertainties inherent in such a forecast.Since even its relation to the respective products is not always clear, the uncertainties of forecasts of a patent evaluation are particularly pronounced. So patent evaluations are dependent on the point in time when they are drawn up and are subject to changes in the course of time, if in the meanwhile the bases for the evaluation have changed. Fig. 4.1 shows an (incomplete) overview of methods for the evaluation of patents along the technology life cycle (Wurzer, 2002).

Option theory

Especially at the beginning of a technological development, the uncertainties about the technical and economic success of an idea are immense. Accordingly, there are numerous options for action in the course of a patent application procedure, both prior to and after a possible patent grant. With this in mind, we may regard patents as so-called real options, and transfer the theory of pricing for purchase and sale options of tradeable goods to the patent evaluation. Here for example, the expected costs and benefits of a possible patent are brought in, but also the expected costs that would arise if a patent were in the hands of a competitor (Pitkethly, 2002).

License analogy

For the method of license analogy, the conditions of license agreements, considered to be analogous with regard to the product market concerned, are consulted for comparison purposes. Hellebrand and Kaube (2001), for instance, issued a compilation of settlement proposals of the arbitration board at the German Patent Office that can be used for the method of license analogy. The application of this method requires a relatively high maturity of the technology.

It has been shown that several bibliometric patent indicators correlate with the value of patents (Wurzer, 2002, Reitzig, 2002 and the references indicated there). They offer the advantage of being reliably collectable from patent databases at manageable costs. However, the use of patent indicators implies that relevant patent information has already been published, hence a certain time lag occurs. In practice, the number of quotations in search reports and third-party patents are used, as well as the number of objections, the number of references in non-patent literature, the existence of appeals and nullity suits plus data about the size of patent families and the legal status of patent applications.

Bibliometric patent indicators

4.1.2 Analysis of Patents on Nanotechnology

In literature no surveys have been found that systematically gather typical patent indicators to evaluate nanotechnology. Nevertheless, there are single surveys on patents in the field of nanotechnology that partly provide important information. A preliminary study on nanotechnology in connection with the European Commission's effort to develop "maps" for technological and scientific excellence in Europe ("mapping excellence") is of special interest (Nanotechnology Expert Group, 2002). Despite the comparatively high quality of this study, the authors themselves state the following restrictions: "It should be noted that all tables reflect only limited patent data" (p. 21). "Abstracts and keywords were missing from much of the patent data" (p. 41). It has to be emphasized that in this study, the patents investigated were presented to experts to assess whether or not the respective patents should be classified as nanotechnology.

Experts' assessment

Further studies deal with the interplay of science and technology in the field of nanotechnology and international cooperative structures, analyzing, inter alia, literature quotations in patents (e.g. Meyer, 1998; Meyer, 2001; Verbeeck, 2002). In this connection, the Hullmann's dissertation (2001) is remarkable, as it examines the international transfer of knowledge, taking nanotechnology as an example, and addresses patents as well.[2)]

Interplay of science and technology

A comparison by country based on data from the US American Patent Office (USPTO) is found in Marinova (2003). In the course of the work for this study, another paper also analyzing data of the USPTO was published (Huang, 2003). While Marinova (2003) establishes a total number of 1524 German patent applications at the USPTO until 2000, in Huang (2003) Germany doesn't even appear as an

2) See also Compaño (2002), based on the same data enquiry with patent data up to approx. 1997/1998.

applying country – a fact that throws doubt on the quality of this work.[3)]

This observation shows, however, that prior to assessing the publication and to drawing conclusions from such a publication, the data basis and the analytical methods have to be carefully scrutinized.

4.1.3
Search Strategy and Methodology

Just as there is no generally accepted definition of nanotechnology (see discussion in Chapter 2), there is also no accepted search strategy for nanotechnology patents. Even the studies mentioned in the previous section differ considerably from each other with respect to the search terms used, nor can an assignment be made on the basis of special classes in the international patent classification (IPC).[4)]

Discussions with representatives of the EPO

Discussions with representatives of the European Patent Office (EPO) during the work on this study contributed to fact that the EPO has become aware of this problem and is now aiming at the development of a new classification on the basis of newly defined ICO codes ("in computer only"). In a first round, a team of about ten patent examiners identified approx. 15 000 patents with reference to nanotechnology[5)].

The multitude of search strategies in literature can be roughly subdivided into the following three groups:

"nano" as prefix

1 The simplest strategy is to search for the prefix "nano", with certain combinations of this prefix being excluded explicitly when there is obviously no connection to nanotechnology. The most frequent of the combinations to be excluded are:
nano-sec? or nano2 or nano3 or nanogram? or nanolite? or nanolitr? or nanomol? or nanos or nanosat? or nanosec? or subnanomol? or subnanosec?

Detailed search word lists

2 A detailed search strategy comprises the compilation of comprehensive lists of search terms. Examples of such lists are to be found in Braun (1997) and Hullmann (2001). In principle, explicit search term lists consist of two elements. On the one hand, there are

3) The background here is probably an error in the country code used, which in Huang (2003) is stated with "DT" for Germany, whereas the correct code is actually "DE".

4) In the IPC, class B82 has the heading "nanotechnology". However, this class subsumes only a few strictly outlined nanostructures and their manufacture.

5) This effort may contribute essentially to the EPO's mastering of the challenge that a new and highly interdisciplinary field, such as nanotechnology, may represent to a patent office. However, ICO codes are not intended to be used by outsiders, so that the difficulties mentioned remain unaffected.

search terms representing nanotechnology in its entirety, if possible (example: "nanoparticle"); on the other hand, there are such search terms that are typical for an individual partial technologies that are regarded as belonging to nanotechnology.[6)]
The difficulty of such rather specific search terms is the compilation of a complete list.[7)] Depending on the partial technologies considered or ignored in the search, distortions and imbalances may arise, especially if the patent assignees are companies with a very specific product and technological profile. On the other hand, with the number of additionally considered partial technologies, there is a growing risk of including search terms that do not only describe nanotechnology exclusively.

3 Consequently, the most thorough strategy is to carry out a wide ranging search of a comprehensive sample of candidates for nanotechnology patents, and to decide whether they actually belong to nanotechnology after an examination of all retrieved patent documents. However, in their survey (2001, p. 47) the Nanotechnology Expert Group and Eurotech Data come to the conclusion: "One cannot expect experts to agree on which (...) are relevant."

Examination of all patents retrived

Due to the large number of patent documents and the limited resources available, a search based on the following detailed list of terms was selected for the purposes of this study:

Search strategy

nanoactuator?, nanoaggregate?, nanoamorphous?, nanoanaly?, nanoarchitectur?, nanoarray?, nanobacteri?, nanoball?, nanoband?, nanobar?, nanobead?, nanobelt?, nanobio?, nanobot?, nanobridge, nanobuildingblock?, nanocage?, nanocapillarity?, nanocapsul?, nanocarrier?, nanocatal?, nanocavity?, nanocell?, nanoceramic?, nanocermet?, nanochannel?, nanocharacter?, nanochem?, nanochip?, nanocluster?, nanocoat?, nanocolloid?, nanocolumn?, nanocomposit?, nanocompound?, nanocomputer?, nanoconduct?, nanocone?, nanoconstriction?, nanoconstruction?, nanocontact?, nanocrack?, nanocrystal?, nanocube?, nanodeformation?, nanodevic?, nanodiamond, nanodiffraction, nanodimension?, nanodisk?, nanodispers?, nanodisplacement?, nanodissection?, nanodomain?, nanodot?, nanodrop?, nanoelectr?, nanoelement?, nanoemulsion?, nanoencapsulat?, nanoengineer?, nanoenvironment?, nanoetching?, nanofabricat?, nanofeature, nanofiber?, nanofibr?, nanofilament?, nanofiller, nanofilm?, nanofilt?, nanofluid?, nanofoam?, nanofriction?, nanogap?, nanogel?, nanoglass, nanograin?, nanogran?, nanogrid?, nanogroove?, nanohardness?, nanoheterostructure?, nanohole, nanohorn?, nanoillumin?, nanoimprint?, nanoimprint?, nanoinclu-

6) Examples and consequences for the analysis of patent data are discussed in the other sections.

7) In principle, search terms for all technologies and applications stated in Chapter 3 would have to be included.

sion?, nanoindentation?, nanoionics?, nanojunction?, nanolaminate?, nanolayer?, nanolithograph?, nanomachin?, nanomagnet?, nanomanipulat?, nanomanufactur?, nanomap?, nanomask?, nanomaterial?, nanomatrix?, nanomechanic?, nanomembrane?, nanomeric?, nanometal?, nanomodification?, nanomolecular?, nanomotor?, nanomultilayer?, nanoobject?, nanooptics?, nanopartic?, nanopattern?, nanophase?, nanophoto?, nanophysics?, nanopigment?, nanopipe?, nanopit?, nanopolar?, nanopolyhedra?, nanopor?, nanoposition?, nanopowder?, nanoprecipitation?, nanoprobe?, nanoprocess?, nanoreact?, nanorheology?, nanorod?, nanoroughness?, nanoscaffolding?, nanoscale?, nanoscien?, nanoshell?, nanosize?, nanosol?, nanosolid?, nanosource?, nanospectroscopy?, nanosphere?, nanostring?, nanostruct?, nanosurface?, nanosuspension?, nanoswitch?, nanosystem?, nanotech?, nanotemplate?, nanotexture?, nanotip?, nanotiter?, nanotool?, nanotopography?, nanotribology?, nanotub?, nanotweezer, nanowear?, nanowelding, nanowhisk?, nanowire?

nanometer(w) (accurac? or partic? or precision? or thick? or thin? or scale? or size? or structure? or width?)

atomic (w) layer(w) (deposit? or epitax?) or molec? (w) beam (w) epitax? or mbe or metal? (w) organ (w) vapo? (w) phase (w) epitax? or movpe

This list has partly been supplemented by further search terms (see explanations in the following sections). Random samples of the patents retrieved have been checked for their nanotechnological relevance. The assignment to a certain branch of industry was carried out using IPC classes based on the OECD concordance (OECD, 1994) and if necessary adding further search terms.

Database WPINDEX

For the patent analysis within the context of the present survey, the database WPINDEX has been chosen. Compared with the studies quoted above, this entails two advantages:

1 WPINDEX is suitable for statistics, as the individual elements of the database each represent complete patent families. This is not the case in other databases, where multiple registrations of the same patent application are possible (e.g. when publications of unexamined applications and granted patents for the same application as well as equivalent patents of the same family are registered as separate documents in a database).
2 The database provider draws up its own short summaries for each patent family. This moderates the consequences of the common practice for patents to deliberately avoid essential keywords (here e.g. the prefix "nano") in the phraseologies of patents.[8]

8) In order to render the search for patents more complicated for competitors, glossing-over is common in patent specifications. So it would not be unusual in a patent, for instance, to describe a "washing machine" as a "device for the cleaning of textile items".

A disadvantage of the WPINDEX database is the fact that it does not allow a search by the addresses of the inventors. For this reason, the assignment of a patent family to a country was carried out on the basis of the country of the priority application (database field PRC "priority country").

Assignment according to countries

The quality of this assignment was verified using the data gathered by the search about the chemical industry: with the top 20 applicants from this field of industry, the country of the priority application corresponded to the country where the head office of the company was situated. The level of agreement reached more than 85% for the various applicants, and in one third of the companies 100% were reached.

In the following sections, patent statistics based on the count of patent families are presented. The analysis evaluates the countries of application, the progression over time of applications and the main topics. In the industry-specific searches, the leading applicants (companies) are also stated.

Count of patent families

Section 4.2 begins with an overview of all nanotechnology patents, followed by analyses for the selected Lead Markets: chemistry in Section 4.3, optics in Section 4.4 and finally car manufacture in Section 4.5.

Moreover, there is an analysis by countries in each section that is restricted to particularly valuable patent families. As an indicator, patent families were chosen for which at least one patent had already been granted[9] and the application has been made in more than one country[10]. This indicator is easily gathered and offers the advantage that statements are possible with relatively little time lag. This form of patent evaluation cannot be converted into monetary values, however, with certain restrictions statements regarding the relative position of different countries in competition are possible.

Particularly valuable patent families

4.2 Patents in Nanotechnology as a Whole

The patents in nanotechnology as a whole were determined with search strategy 2 on the basis of the comprehensive word list given in Section 4.1.3.[11] Altogether, about 13 000 patent families were found.[12]

9) According to the database field PK (patent kind code).

10) According to the database field CYC, Criterion CYC >1.

11) Date of the search: March 4, 2004.

12) In the following, the terms patent family and patent are used synonymously.

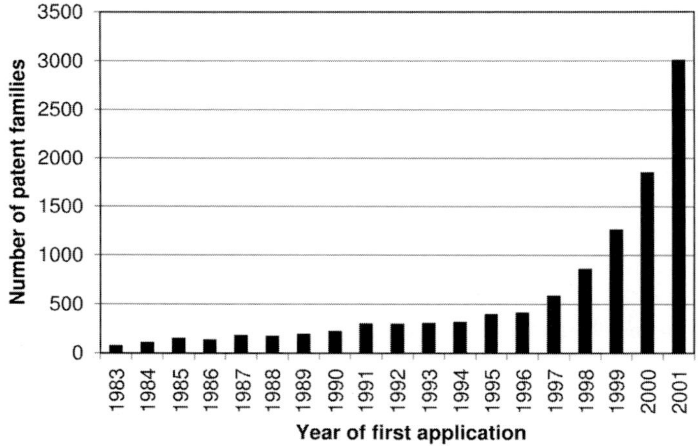

Figure 4.2 Increase in number of patent applications over time in nanotechnology as a whole.

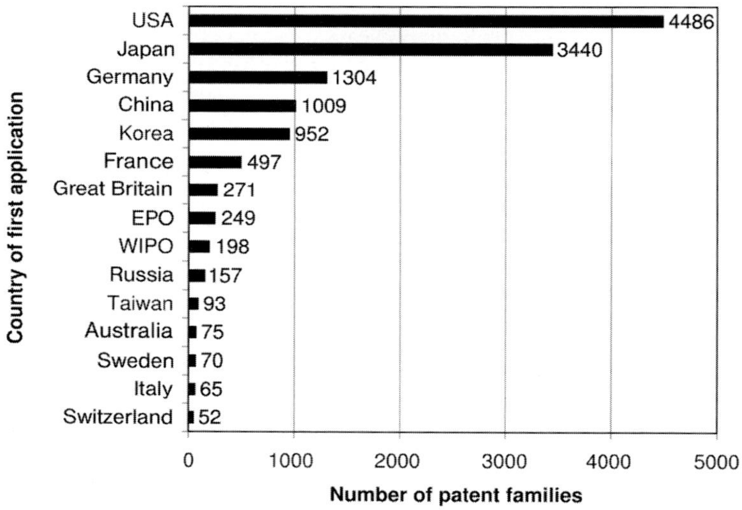

Figure 4.3 Country distribution of patent applications in nanotechnology altogether.

Figure 4.2 shows the increase in number of the applications over time. A jump in the applications since the end of the 1990s is glaringly obvious, so the number of applications has more than doubled in the periods from 1997 to 1999 and from 1999 to 2001[13].

Most patents are held in the USA with a number of almost 4500, followed by Japan with 3440 patents and Germany with approx. 1300 patents (see Fig. 4.3).

13) Since the time delay between the application for a patent and the first publication a database is able to record, can be 18 months, no complete figures are currently available for the years after 2001. Thus, for the sake of clarity, these years are not shown in the charts of the time progression. The other charts, however, already contain available data from 2002, 2003 and 2004.

Surprisingly, China and Korea[14] with about 1000 patents each follow immediately, even before France with approx. 500 patents and Great Britain with nearly 300 applications. Under the category EPO ("European Patent Office") and WIPO ("World Intellectual Property Organization") applications are subsumed that have not been filed in one of the national patent offices, but directly in one of these two supranational patent offices.[15] Russia, Taiwan, Australia, Sweden, Italy and Switzerland follow.

To evaluate this ranking order and assess the relative position of the leading countries, some weighting factors must be taken into account. According to statements of EPO experts, the quality and topical scope of individual patents should be taken into account in country-by-country comparisons. There is a difference in quality and scope between European applications, i.e. also in particular German patents, and applications from the USA and Japan. According to the experts of the EPO, it is therefore on average suitable to regard one European or German application as equivalent to 7–9 Japanese and 3–5 USA applications. In consideration of these weighting factors, Germany lies at least equal with the USA and significantly ahead of Japan.

Weighting factors for country-by-country comparisons

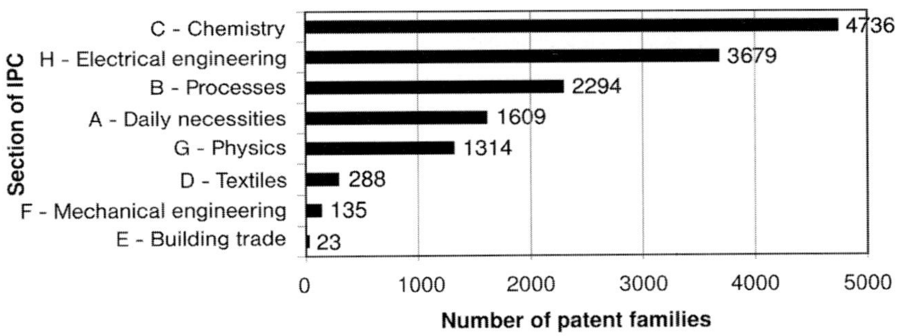

Figure 4.4 Distribution of all nanotechnology patents with respect to the IPC classes.

Figure 4.4 shows the rough thematic distribution of nanotechnology patents on the basis of their relation to the eight sections of the IPC. The largest number of patents, i.e. good one third of all nanotechnology patents, is assigned to Section C "Chemistry and Metallurgy", which includes the sector biotechnology. A good quarter of nanotechnology patents derive from the section of electrical engineering. Patents from Section B "Working Processes, Transporting" are third.

Chemistry ranking first

14) Here and in the following, the name "Korea" is used synonymously for the Republic of Korea ("South Korea").

15) Owing to the above-mentioned limitations of the WPINDEX database, a more precise country distribution is not possible for these applications.

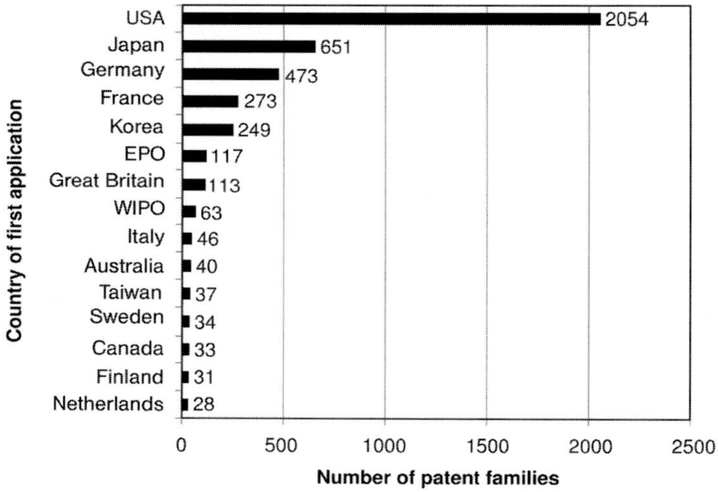

Figure 4.5 Country distribution of patent families with at least one granted patent and with applications made in more than one country, for nanotechnology as a whole.

As well as the whole process engineering this section contains surface and coating technologies, too. Section A, "Everyday Necessities", holds one eighth of all nanotechnology patents. In particular, this section includes cosmetics and medical technology. About 10% of nanotechnology patents belong to Section G "Physics". The other sections like "Textiles, Paper", "Construction Industry; Mining; Drilling" and "Mechanical Engineering; Lighting; Heating; Arms; Blasting" are only of secondary importance with regard to nanotechnology patents.

Particularly valuable patents

Figure 4.5 shows the country distribution of those patent families with at least one granted patent[16] and where the application has been made in more than one country. According to these statistics, the USA again holds most of the patents, followed by Japan and Germany. Looking at the number of these patents compared to all patents, it is noticeable that among these three countries Japan shows the lowest quota with less than 20% (see Fig. 4.3). If these weighting factors are taken into consideration, the USA and Germany would then again be more or less equal.

Compared with Fig. 4.3 it is noticeable that China no longer appears among the top 15 countries, although it ranks fourth in the count of all patent families. However, application figures for China have only

16) Search for the following patent class codes: aub or ca# or cha or cha5 or chb# or cnc or dec# or dee or deg or epb# or fib1 or frb or gbb or itb or jpb# or krb# or nlc# or ruc# or sec# or usa or usb# or usc# or use#.

been able to be recorded since the year 2000, so it might be possible that with regard to the applications from China, the time lag for the indicator chosen here is still insufficient. Now France appears in fifth place with a comparatively high quota of 55%, before Korea with a quota of only about 25%. Then Great Britain, Italy, Australia, Taiwan, Sweden, Canada, Finland and the Netherlands follow.

4.3 Nanotechnology Patents in the Field of Chemistry

This section presents the results of the search for nanotechnology patents in the field of chemistry.[17] The assignment of the patents to the field of chemistry was carried out on the basis of the OECD concordance (OECD, 1994, P. 77). The fields of macromolecular chemistry and polymers, materials and metallurgy, organic fine chemistry and basic chemistry were taken into account.[18]

Most of the patents come from the USA, followed by Japan, (see Fig. 4.6). In the third place, applications from China already appear, and then applications from Germany follow before France and Korea.

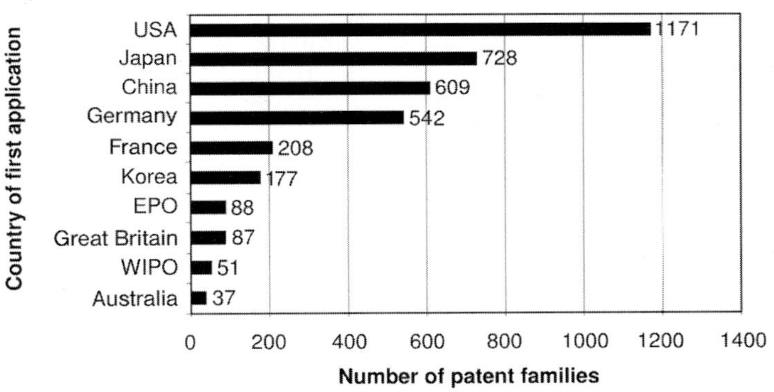

Figure 4.6 Country distribution of nanotechnology patents in the field of chemistry.

17) Date of the search: June 18, 2003. This search was carried out according to search strategy 1 from section 4.1.3. A comparison with search strategy 2 has shown that in individual cases deviations of around 20 to 30% arise. However, the qualitative results do not change considerably.

18) This corresponds to the following IPC-classes: (c07c or c07d or c07f or c07h or c07j or c07k or c08b or c08f or c08g or c08h or c08k or c08l or c09d or c09j or c01! or c03c or c04! or c21! or b22! or a01n or c05! or c07b or c08c c09b or c09c or c09f or c09g or c09h or c09k or c10b or c10c or c10f or c10g or c10h or c10j or c10k or c10l or c10m or c11b or c11c or c11d)/ic.

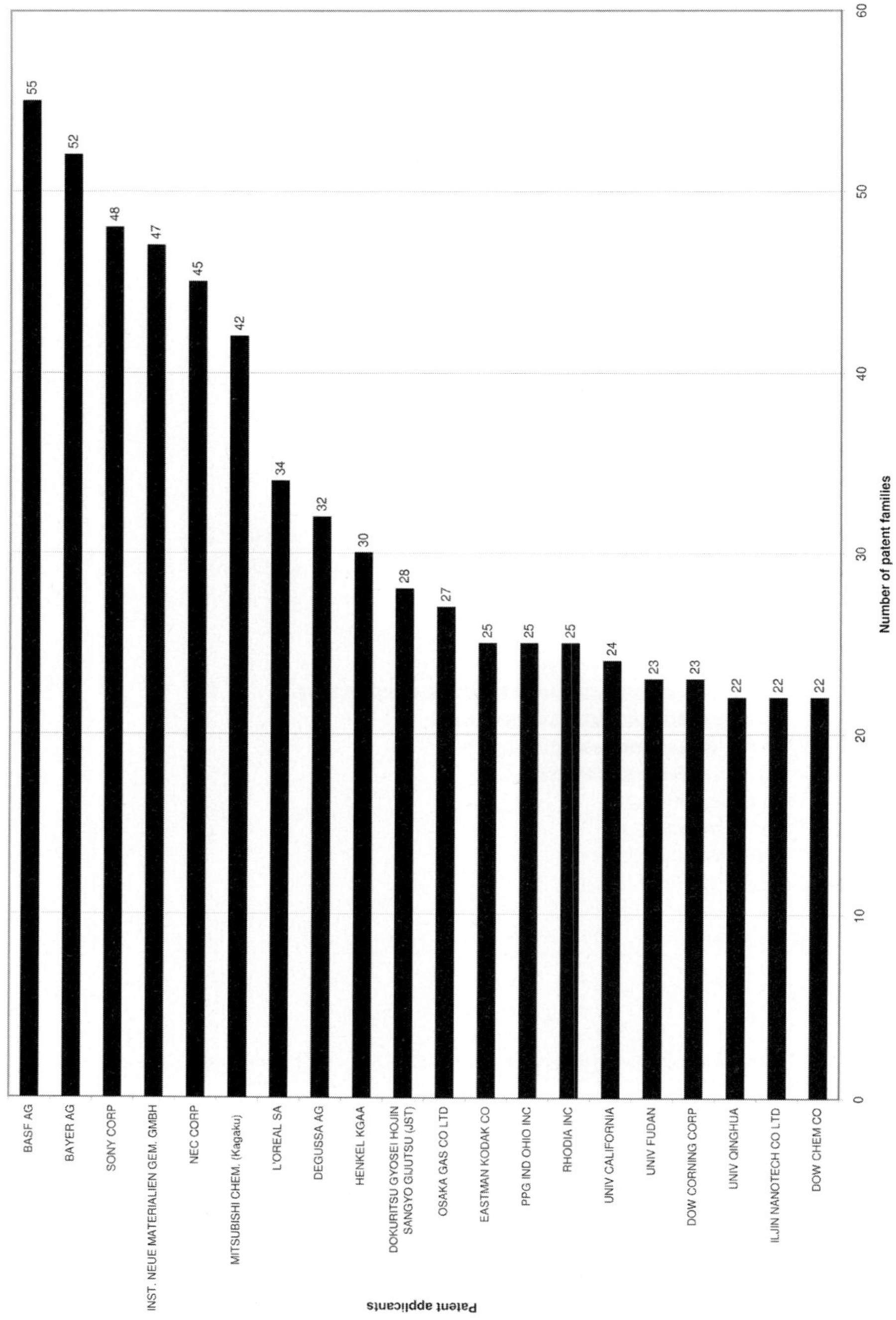

Figure 4.7 Leading applicants for nanotechnology patents in the field of chemistry.

Apart from the supranational applications, Great Britain and Australia appear among the top ten.

Taking the above-mentioned weighting factors into account again, Germany is slightly ahead of the USA and clearly in front of Japan. In the field of chemistry, the high number of applications, especially from China, but also from Korea is conspicuous as well.

The list of the top 20 applicants is shown in Fig. 4.7. Among the ten leading applicants, five are from Germany alone, these are BASF, Bayer, Degussa, Henkel and the INM, Saarbrücken.

German companies in good position

During the presentation of the results before company representatives and industry experts within the framework of a workshop (see Section 6.3) it became clear that with this search, only about one third to one quarter of all relevant patents were found. Nevertheless it can be assumed that the ratios are described more or less exactly.

On the other hand, this compilation verifies the problem discussed initially of considering specific individual technologies; e.g. according to Covion's own information, this company holds about 70 patent families with relation to organic semiconductors, a technology that has certainly to be regarded as belonging to nanotechnology. Great effort seems to be required to establish a search strategy that considers all special suppliers adequately in this respect. Therefore, a restriction has to be made by saying that application figures of companies with very specific profiles are probably under-represented.

Example Covion – special suppliers difficult to gather

Keeping this restriction in mind it can be said that the leading Japanese applicants are Sony, NEC, Kagaku, the JST and Osaka Gas. As applicants from France, L'Oreal and Rhodia appear. American applicants are PPB Industries, Eastman Kodak, Dow Corning and Dow Chemical as well as the University of California. Leaders from China are the Universities of Fudan and Qinghua.

The topical priorities are depicted in Fig. 4.8. The largest number of applications comes from the field of materials, even before the applications from polymer chemistry. Taking the increase in number of applications over time into account, these two fields currently show the sharpest increase. The number of applications in the field of chemistry as a whole qualitatively goes up just as it does in the case of nanotechnology patents as a whole.

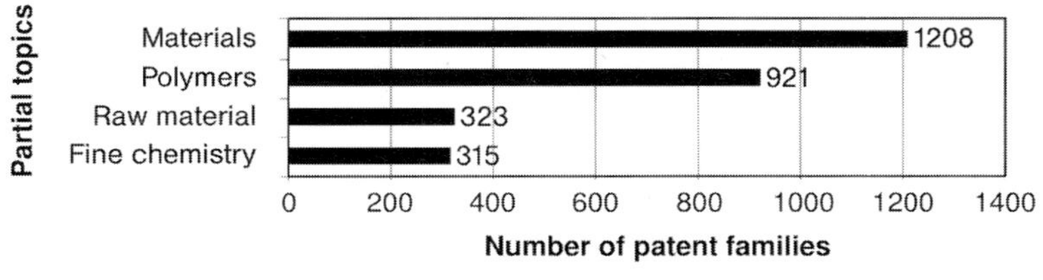

Figure 4.8 Topical focus of nanotechnology patents in the field of chemistry.

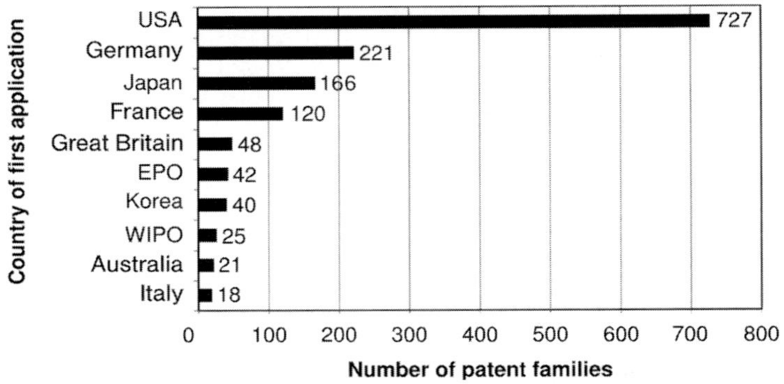

Figure 4.9 Country distribution of patent families with at least one patent granted and with applications made in more than one country for nanotechnology patents in the field of chemistry.

Particularly valuable patents

Figure 4.9 shows the country distribution of patent families with at least one patent granted[19] and with applications made in more than one country.[20] In these statistics, the USA holds most of the patents, followed by Germany and Japan. From these three countries, it is again Japan that has the lowest quota (ratio of the number of these patents to all patents, according to Fig. 4.6) with less than 25%. Considering the weighting factors mentioned above, the USA and Germany would again be almost equal, however, both clearly ahead of Japan.

In comparison with Fig. 4.6 it is particularly noticeable that China no longer appears among the top ten applying countries in these statistics, although it has still been third after the count of all patent families.

France is now fourth with a comparably high quota of 60% before Great Britain, Korea, Australia and Italy. The quota of Korea is less than 25%.

4.4 Nanotechnology Patents in the Field of Optics

Search strategy nanooptics

The search concerning nanotechnology patents in the field of optics was carried out according to search strategy 2 from Section 4.1.3, supplemented by the following keywords that are regarded as relevant for the whole of nanooptics:

19) Search for the following patent kind codes: aub or cnc or dec# or dee or deg or epb# or frb or gbb or jpb# or krb# or usa or usb# or usc# or use#.

20) Criterion cyc>1.

quantumcascad? or quantumdot? or quantumfilm? or quantum-heterostructure? or quantumpot? or quantumwell? or quantumwire?

optic?(w)near(w)field? or snom or near(w)field?(w)optic? or snim or scanning(w)near(w)field(w)infrared

surface(w)plasmon(w)resonan?

fluorescen?(w)resonan?(w)energ?(w)transfer? or fret or single(w) molecule(w) (spectroscop? or fluorescen?) or fluoresc?(w)correl? (w)spectrosc?

(femtosecond or ultrashort(w)pulse)(w)laser and (cut? or drill? or ablat?)

vertic?(w)cavit?(w)surfac?(w)emi?(w)laser? or vcsel

ultraprecision(w)optic?

The assignment of the patents to the field of optics was carried out following the OECD concordance (OECD, 1994, P. 77). The technological field of "Optics", as it is called there, was taken into account supplemented by certain IPC classes, inter alia regarding lighting and electric light sources, non-visible wave lengths and ultraprecision processing of optical elements.[21]

Figure 4.10 shows the distribution of the patent applications with respect to the countries of the priority application.

In the field of optics, most of the nanotechnology applications are from Japan, still before the USA. At a considerable distance, Korea follows before Germany, Great Britain and France. Taiwan and China exhibit clearly less than 100 applications.

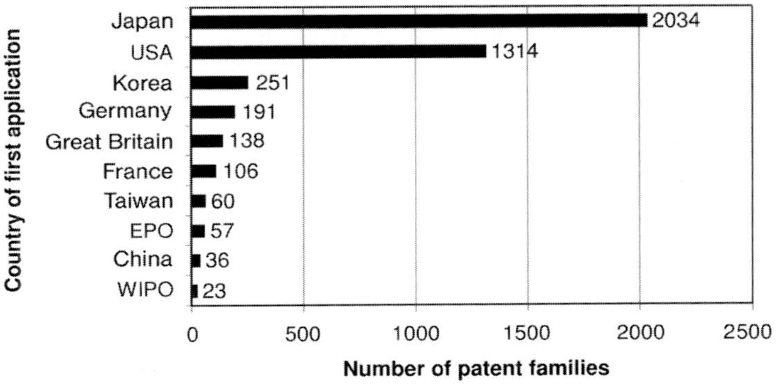

Figure 4.10 Country distribution of nanotechnology patents in the field of optics.

21) Date of the search: October 24, 2003. Restriction regarding the IPC classification: (g02! or g03! or h01s or b24b013 or b29d011 or f21! or g01m011 or h01j or h01k or h01l033 or h05b031 or h05b033 or h05b035 or g01j)/icm.

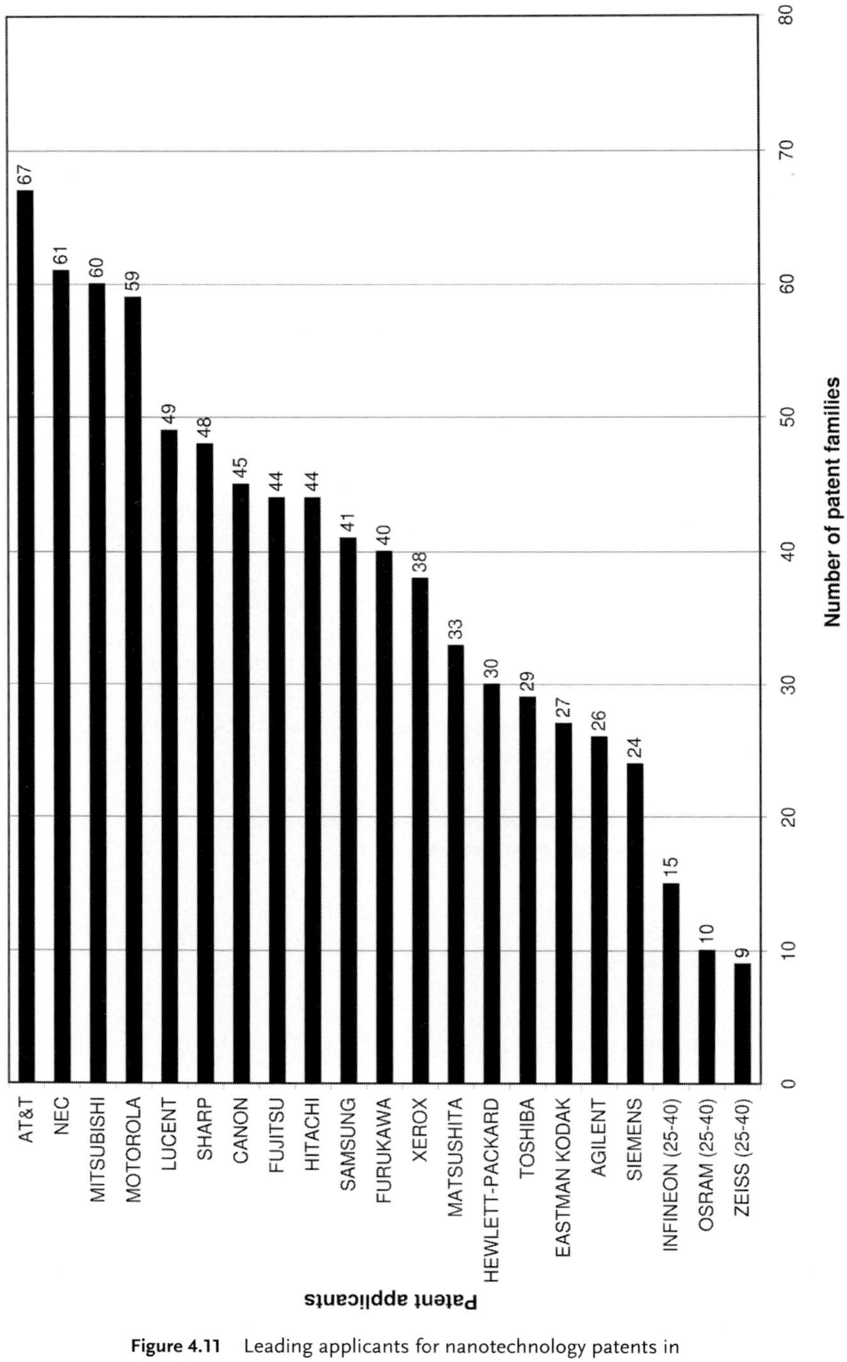

Figure 4.11 Leading applicants for nanotechnology patents in the field of optics.

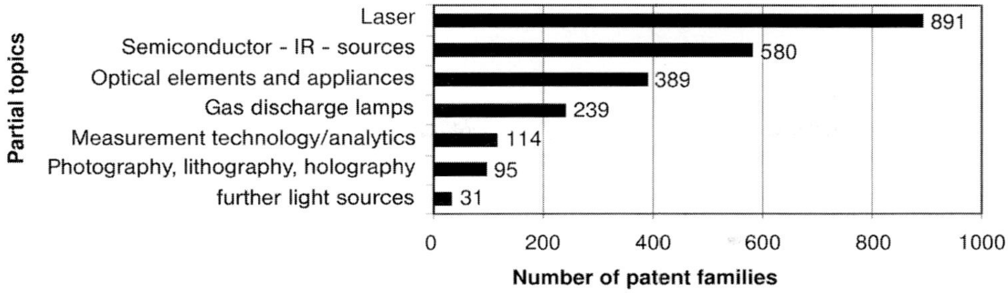

Figure 4.12 Topical priorities of nanotechnology patents in the field of optics.

The list of applicants according to the country distribution is led by American and Japanese companies (Fig. 4.11). Among the top 20 applicants, the only German company appearing here is Siemens. Other German companies among the top 40 are Infineon, Osram and Zeiss.

Figure 4.12 shows the topical priorities in nanooptics. All in all, light sources prove to have the largest scope. The two largest sectional topics are nanooptical lasers and semiconductor-based infrared sources. Moreover, in fourth place gas discharge lamps follow. Other optical elements and devices make up approx. one sixth of all applications in nanooptics. Further section topics are measurement technology and analysis as well as photography, lithography and holography.

The increase in number of the applications in the field of optics is shown in Fig. 4.13. This differs from the increase in number of the ap-

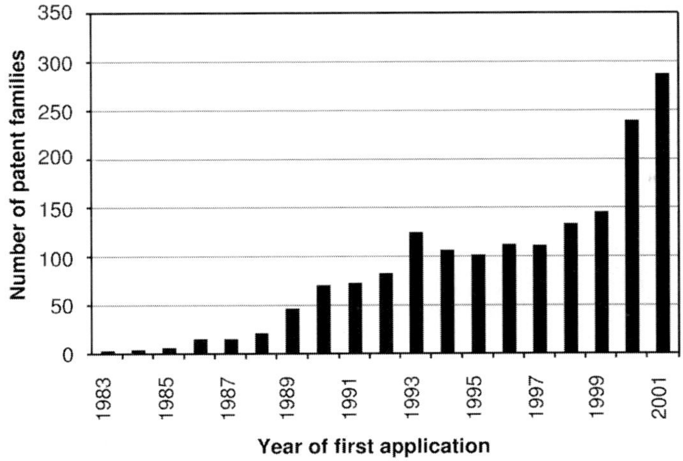

Figure 4.13 Increase in number of applications over time for nanotechnology patents in the field of optics.

plications in nanotechnology as a whole, with a first peak of applications in the year 1993. Upon examination of the increase in number of the applications in the different partial topics it becomes clear that this peak goes back to the field of lasers and semiconductor-based sources. Here it is evident that with the sources based on quantum films, one specific nanotechnology became important early on – even economically. In this respect, the overall picture appears to be the total of the applications to light sources based on quantum films, with a peak in the early 1990s and the trend, noticed for nanotechnology as a whole, with a gradual increase of patents and a sharp increase since the end of the 1990s.

Quantum film laser as an early application

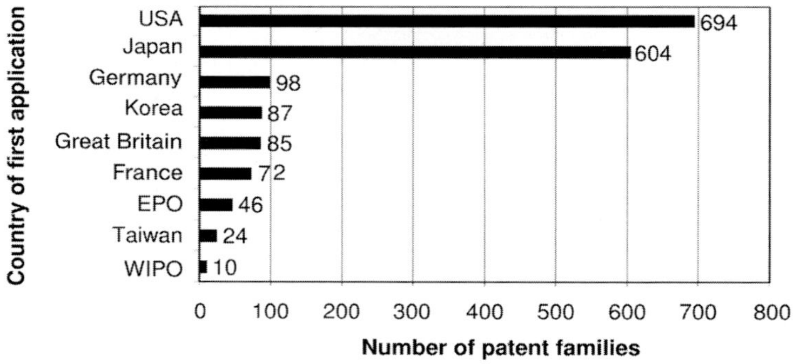

Figure 4.14 Country distribution of patent families with at least on patent granted and with applications made in more than one country, for nanotechnology patents in the field of optics.

Figure 4.14 shows the country distribution of patent families with at least one patent granted and with applications made in more than one country[22]. In these statistics, the USA ranks before Japan. In absolute figures, the difference of these two countries from Germany is considerable. However, taking into consideration the weighting factors discussed above, Germany and Japan seem to be almost equal. Then, Germany's distance, but also the distance of Great Britain and France to the USA, is clearly smaller. Korea ranks fourth; with approx. 35% in the field of optics, Korea's ratio of the number of these patents to the number of all patents (see Fig. 4.10) is significantly higher than Korea's average in nanotechnology as a whole.

Particularly valuable patents

22) see footnotes 19 and 20.

4.4.1
Lithography with Extreme Ultraviolet Light (EUVL)

In Fig. 4.15, the increase in number of the applications in a special search regarding lithography with extreme ultraviolet light (EUVL) is depicted[23].

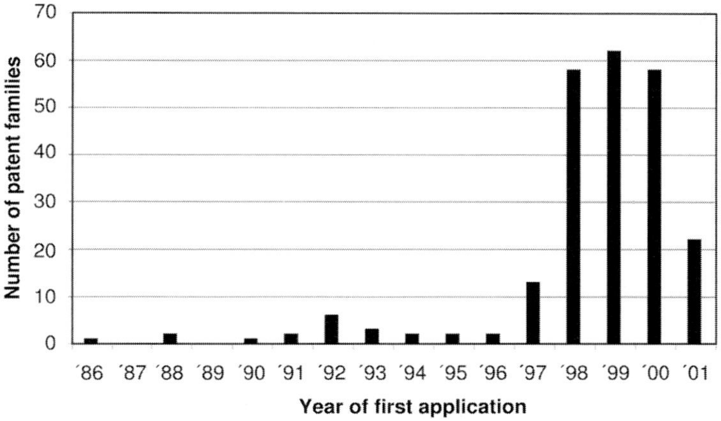

Figure 4.15 Increase in number of patent applications over time in the field of EUVL.

Here, there is a clear jump in activities in the year 1998, attributable to the worldwide intensified effort to develop EUVL[24].

On examination of the companies important for steppers, Fig. 4.16 shows a clear superiority of the ASML-SVGL-Zeiss Group. Nikon has a good position just before the EUVLLC, whereas Canon has only very few applications.

This example of EUVL proves that evaluation of special technologies also requires specific patent analyses. So in the general analysis (see Fig. 4.11) Zeiss does not appear in an outstanding position; its excellent technology position in the economically important EUVL technology becomes clear only through a specific search. Therefore, further detailed searches would be necessary for a complete evaluation of Germany's technology position in the field of nanooptics.

Complete evaluation of the technology position requires detailed searches

23) Date of the search: February 1, 2002. Search for: (euv? or extreme(1w)ultraviolet?)/bi and (g03f? or h01l021?)/icm.

24) The data for 2001 was still incomplete at the time of the search.

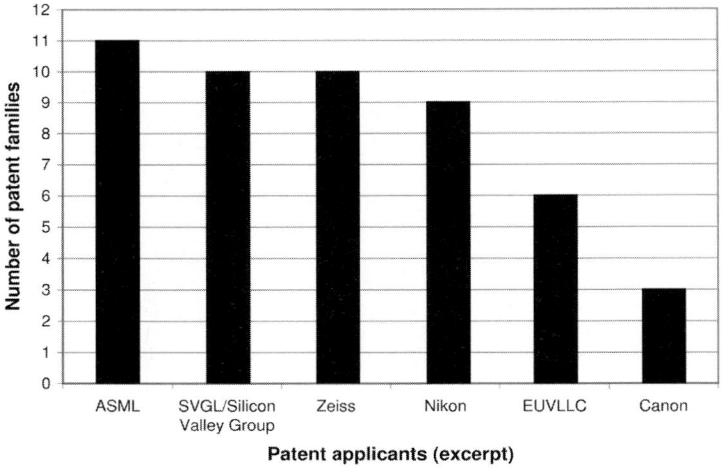

Figure 4.16 Extract of leading applicants in the field of EUVL.

4.5 Nanotechnology Patents in the Field of Car Manufacture

Search strategies in car manufacture

In this section, the search regarding nanotechnology patents in car manufacture is presented. The search was carried out according to strategy 2 (comp. Section 4.1.3). There are two classes in the IPC, B60 and B62, that are dedicated especially to the automotive industry and motor vehicles. The preliminary search, however, shows that the number of nanotechnology patents in these classes is very low. Therefore, the search was carried out without restrictions on the IPC classes, and the assignment to automotive engineering was attempted via the additional search terms CAR OR VEHICL? OR AUTOMOB? OR AUTOMOT? [25]

The titles of the remaining set of 238 patents were completely checked for relevance to nanotechnology and automotive engineering. The small number of patents alone, however, indicates that this search is certainly not complete.

Relevance to car manufacture often cannot be derived from patent documents

During the presentation of the results before industry experts within the framework of a workshop (Chapter 6.4), the incompleteness of this search was rightly pointed out. In particular, it was emphasized that the relevance of many patents to the automotive industry often cannot be derived from the patent documents. This concerns mainly

25) Date of search: November 17, 2003. Owing to the search word "vehicl?" the hit rate showed a relatively large portion of patents from the field of biotechnology/pharmacy where terms such as "drug delivery vehicle" are of some importance. Therefore, the search words: drug# or gene? or agent# or protein# or enzym? were excluded.

chemical products used in car manufacture. If, for example, a nanofiller is used in order to enhance specific properties of synthetic materials or lacquers finally also applied in the car sector, the respective patent usually does not contain any reference to this.

In the end, this reflects that the car industry is positioned more at the end of the value-added chain, and its connection to the fundamental basic technologies, such as nanotechnology, exists only via different intermediate stages. Since this fundamental problem can only be solved with great effort, these search results are nonetheless presented very briefly and with the explicit reference to their very limited completeness, and thus to their very limited representative nature.

Search results very incomplete

Figure 4.17 depicts the country distribution that is basically showing characteristics comparable to the country distribution presented above, in which the USA, Germany and Japan have most of the applications, but altogether with a significantly lower total number.

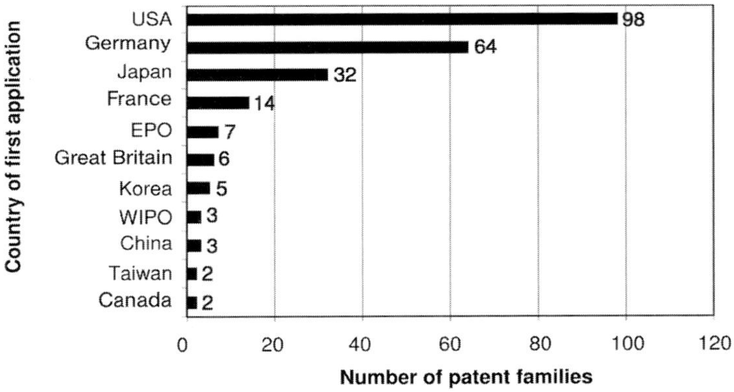

Figure 4.17 Country distribution of nanotechnology patents in the field of car manufacture.

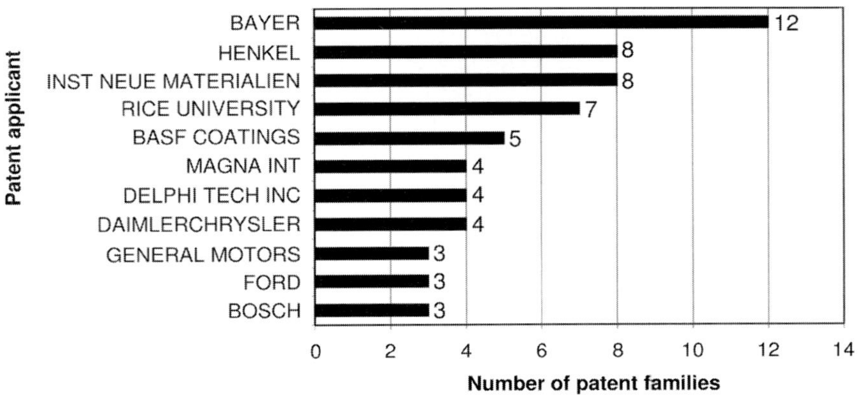

Figure 4.18 Applicants for nanotechnology patents in the field of car manufacture.

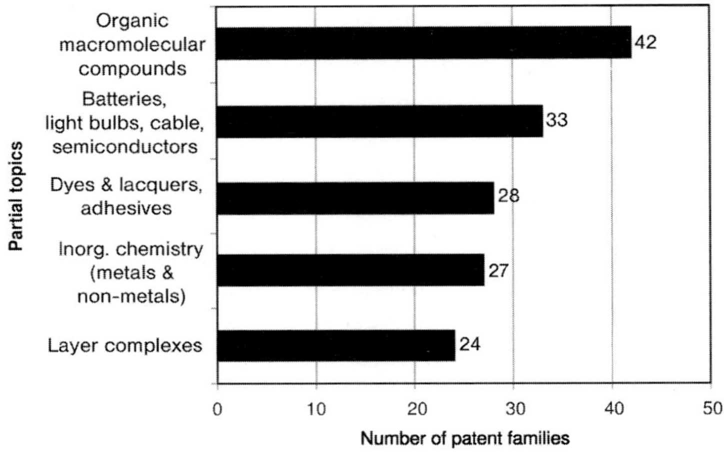

Figure 4.19 Topical focus of nanotechnology patents in the field of car manufacture.

The list of applicants given in Fig. 4.18 begins with companies from the field of chemistry. Henkel, for instance, is active in the field of adhesives for the car industry. Then, with Magna and Delphi, suppliers are following before car manufacturers finally appear.

Important product groups are represented

Figure 4.19 shows the thematic distribution in the retrieved set of patents based on the analysis of the distribution of the IPC classes. Apart from the topical fields given in Fig. 4.19, there are: optical elements and appliances, processes for the application of liquids, coating of and with metals, physical and chemical processes, processing of plastics, chemical generation of fibers as well as vehicle parts, drives and windows. It is noticeable in a positive way that despite of the small size of the retrieved set of patents, those product groups and fields of application of nanotechnology considered to be important for car manufacture are included.

The increase in number of the applications given in Fig. 4.20 also reflects the position of car manufacturing at the end of the value-added chain. It can be seen that the development, compared to the progression of nanotechnology as a whole, begins about 5 years later (see Fig. 4.2).

On examination of patent families with at least one patent granted and with applications made in more than one country[26], only the four countries depicted in Fig. 4.21 remain with more than five applications, so that the number of cases is too small to make more far-reaching statements.

26) See footnotes 19 and 20.

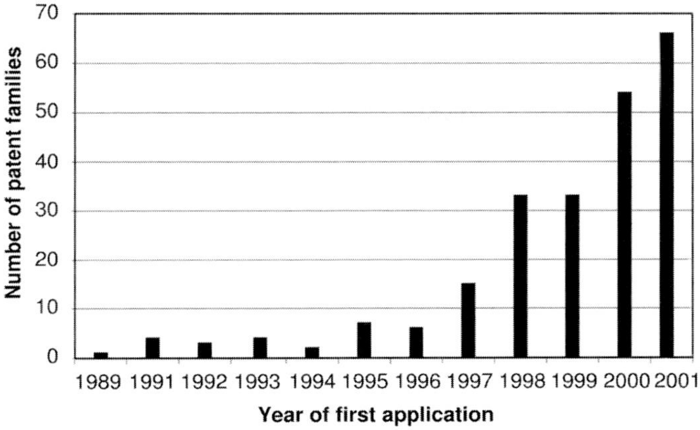

Figure 4.20 Increase in number of applications of nanotechnology patents over time in the field of car manufacture.

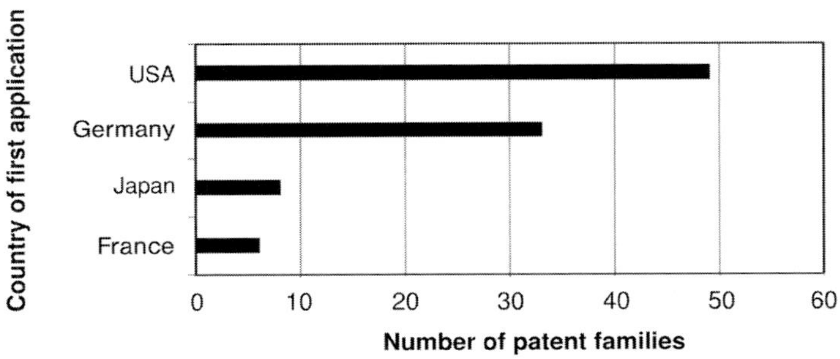

Figure 4.21 Country distribution of patent families with at least one patent granted and with applications made in more than one country for nanotechnology patents in the field of car manufacture

4.6 Nanotechnology Patents in the Field of Medicine and Life Sciences

Independent study on nanotechnology and health

As consortium leader on behalf of the BMBF, the Aachener Gesellschaft für Innovation und Technologietransfer (AGIT) is working on a study on the subject of nanotechnology and health. A patent analysis in the field of medicine and life sciences will be found when the final report of this study, dedicated in detail to this complex topic is published.

The general problems regarding the search strategy for nanotechnology patents is exemplified by the limitations shown in the fields of

High precision of search to be reached only with great effort

medical technology and life sciences. As shown in the technology analysis "Nanotechnology II" (Wagner, 2004) for instance, nanotechnology is applied in certain forms of biochips. For the search words (dna? or bio? or protein?)(w)chip? alone, the WPINDEX database contains more than 900 patent families, only a fraction of which should really be assigned to nanobiotechnology in the literal sense. However, the precise depiction of this fraction by means of search words is extremely difficult. This example makes clear that attempting to describe nanotechnology with the help of a list of individual technologies contains particular pitfalls which may lead to search results of only little precision.

4.7 Overview of Nanotechnology Patents

Dynamic development for patent applications in nanotechnology

Evaluation of the patent situation firmly establishes that the extremely dynamic development of nanotechnology is reflected in patent applications. During the past five years, the annual number of patent applications in nanotechnology has doubled about every two years. (see Fig. 4.2).

Germany very well positioned

Analysis of figures for patent applications across different countries, as presented above, shows that Germany is in an excellent position, both in nanotechnology as a whole, and in the quantitatively most important section of chemistry. (see Figs 4.5, 4.9). Germany's position appears to be particularly good regarding the especially valuable patents. In none of the Lead Markets (chemistry, car manufacture, and optics) that were surveyed in the patent analysis is Germany lagging seriously behind the USA or Japan.

Long development times

In the work-shop discussions on the patent strategies in the field of nanotechnology (see Chapter 6), industry experts agree that, in principle, nanotechnology does not require completely different patent strategies in comparison to other fields of technology.[27] A characteristic, however, is the fact that in certain cases the period of development from a nanotechnological basic effect to the application can be so long that patent protection expires shortly after the maturity stage of the product has been reached.

27) The interdisciplinarity of nanotechnology confronts patent offices with difficult tasks, as experts of the EPO emphasized. So officers must ensure that patent applications are assigned to appropriately qualified patent examiners. This may require the examination of individual applications by a team. They also have to avoid accepting inadmissibly broadly formulated patent claims from lack of detailed knowledge.

Industry experts point out that in small and medium-sized enterprises (SME) the tendency to patent is far more modest than in large enterprises. Should the situation arise, SME prefer not to disclose technological trade secrets at all – even if they were patentable. For financial reasons, SME are more likely to register utility patents, for example. According to the experts canvassed, a reasonable patent strategy for SME could be to have a technology patented together with the main users.

Patent activities of SME's

Finally, it should be emphasized once more what has already been shown in the course of the Chapter in various examples (like in Section 4.4 based on patents for lithography with EUVL in Figs 4.15 and 4.16). We must emphasize that a search with such a wide topical scope as that presented here, is only appropriate for a general assessment of the patent situation. The assessment of individual companies and specific technologies requires more detailed patent searches.

Broad search for the assessment of the general patent situation

Assessment of specific technologies requires specific searches

As set out in Section 4.1.1, analysis of bibliometric patent indicators cannot provide an assessment of the market potentials in absolute figures. However, it becomes clear that considerable efforts to secure intellectual property in the field of nanotechnology are in full swing worldwide – a fact that suggests that patent applicants see a significant market potential in this field. As the analysis shows, Germany has an excellent position in the international competition with regard to the patent situation.

No assessment of market potentials in absolute figures

5
Incorporation and Realization of Nanotechnology in German Enterprises

Thomas Heimer, Hermann Sanders and Norbert Malanowski

5.1
Aims of the Company Survey

Nanotechnology is considered to be one of the pioneering key technologies of the future. Nevertheless, the full scale of the manifold possibilities of nanotechnology is not nearly known yet. On the contrary, only the first indications of innovations through nanotechnology have so far been discovered. These are applied, for instance, in carbon nanotubes, in the reading heads of hard disks of computers or in biochips. The interesting thing about it is the fact that nanotechnology does not simply represent a smaller scaling than microtechnology, but that the reduction in size permits new physical, chemical and/or biological properties to occur that release the potential for completely new approaches to solutions with regard to the economic utilization.

New approaches to solutions regarding economic utilization

Against this backdrop, a paper-based company survey was carried out (see Appendix 2) in order to identify and evaluate the economic potential of nanotechnology in Germany. The main questions of the company survey were:

Company survey of economic potential

- Where is nanotechnology already used today?
- Which growth potential will the economic players give nanotechnology in the years to come?
- Which relations already exist between economic players in nanotechnology today?

In order to achieve standardization of the answers, the whole company survey was based on a uniform definition of nanotechnology (see more detailed discussion in Chapter 2).

Here, nanotechnology is defined as:

Definition

a) all products with at least one functional component with a controlled geometrical size below 100 nanometers in at least one directional dimension rendering physical/chemical or biological effects usable that do not occur above this critical size;

Growth Market Nanotechnology. Edited by N. Malanowski, T. Heimer, W. Luther, and M. Werner

ISBN: 3-527-31457-1

b) analytical and/or process engineering equipment required for the controlled manufacturing, positioning or measurement of the functional components mentioned under (a).

The questions quoted above make clear that it is less about the gathering of future results of basic research than about the chances of diffusion of the existing basic knowledge into research and development in the companies and their transfer into economic applications. Therefore, the object of knowledge is not research, but the economic use of nanotechnology.

5.2
Notes on the Procedure

The questionnaire was formulated and the questions selected to ensure that the companies reported three time spans with regard to nanotechnology:

- the behavior of the companies in the past;
- the present procedure in the companies;
- the strategic orientation of the companies.

Moreover, the questions were worded to focus on the companies' own activities in R&D and in the product range, as well as networking in the supply of nanotechnological components and the supply of such components itself (see Appendix 2).

Selection of the companies polled

For the quantitative survey, the partners participating in the project collected addresses. Future Technologies Division of the VDI Technology Center GmbH supplied 509 selected addresses of companies in Germany that were clearly involved in nanotechnology. A comparison with several later publications revealed that these addresses contain all the companies dealing with nanotechnology that others had listed. Comparison with the data stock of the Deutsche Bank AG showed that the 78 companies the Deutsche Bank AG assigned to nanotechnology, are all included in the VDI TZ ZTC material. For numerous companies, in particular large-scale enterprises, several addresses were available. Altogether 73 addresses were furnished for 25 companies; Siemens AG alone shows eight addresses. This survey was specifically aimed at companies only, so scientific institutes were not included.

Deutsche Bank contributed an additional 291 addresses, in this case of companies active in Microsystems Technology (MST) at the interface with nanotechnology. In this material, the number of addresses is again higher than the number of companies: for 13 companies, two addresses each were stated. In this parent material, about ten engineering offices are listed that, like scientific institutes, were not intended as the target group of the questionnaire campaign.

Therefore, it can be assumed that instead of 800, there is a total of 700 addresses of companies that might be active in the field of nanotechnology, always supposing that this is the case with the approx. 270 company addresses of the Deutsche Bank. If this address material were disregarded, the parent material would comprise about 450 addresses of companies that are sure, or at least suspected, to be active in the field of nano.

Altogether 800 questionnaires were sent out. After the due date had passed, about 70 questionnaires has been completed and sent back, however, the returns came almost exclusively from the VDI TZ ZTC stock. The returns from the MST address material amounted to only five completed questionnaires at that time. In the first round, at least 26 companies that had not sent back a completed questionnaire, reacted and explained their reasons for it. 18 out of 26 explained their failure to answer with the fact that they were not working in the field of nanotechnology, 15 of these addresses came from the VDI TZ ZTC.

After a follow-up campaign by telephone and electronic means, the returns increased to 107 completed questionnaires, with two questionnaires not being included in the quantitative analysis owing to their late mailing.

Return rate

Depending on the figure the parent material is based on, and on the evaluation of the returns, different return rates result. The margin ranges from 13.1 for 105 completed and processed questionnaires in relation to 800 questionnaires sent out, to 15.0 (700 company addresses) and 27.1 for 122 (107 plus 15 "no activity in the field of nano"), in relation to the approx. 450 company addresses of the VDI TZ ZTC parent material.

About 450 companies altogether

In any case, the return rate is sufficient for achieving secure and representative statements with the help of quantitative processing.

The analysis of the returns was carried out with the SPSS program, as was the calculation of the corresponding correlations.

5.3
Structural Data of the Companies

In order to be able to form an opinion about the sample, it is reasonable to gather some information about the structure of the companies polled. The majority of the companies polled, about 70%, was founded in the period between 1980–2002 (Fig. 5.1).

Start-up wave in the 1990s

Obviously, a strong start-up wave in the field of nanotechnology took place in the 1990s. However, this must not be mistaken for the economic power of the companies. Companies founded before 1980

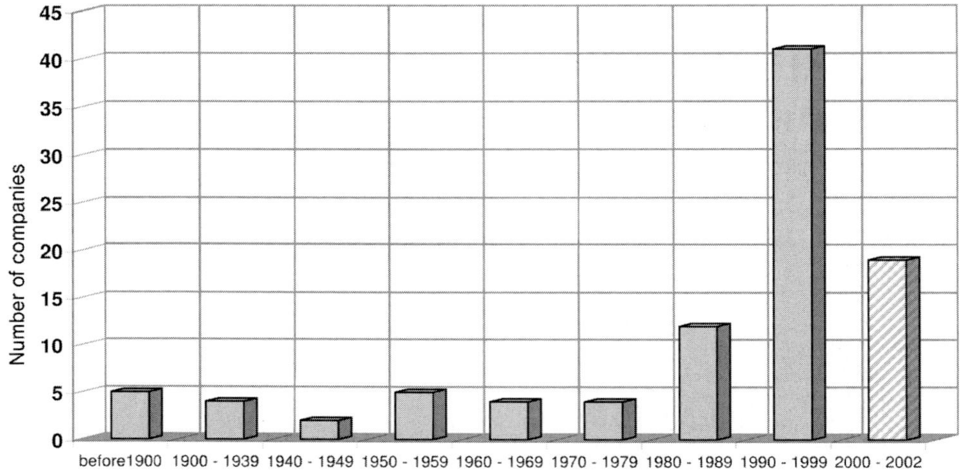

Figure 5.1 Time of business start-up of the companies polled.

that were economically powerful in nanotechnology are, for example, Bosch and Siemens. Even in the case of younger companies, the time of business start-up does not say anything about their economic power.

Economic independence of the companies

An interesting criterion regarding the economic potentials of the companies polled is the issue of economic independence. The question is mainly about whether the company concerned is able to act independently in its location or whether it is an "extended workbench" of a foreign enterprise. The survey shows that 71% of the companies are economically independent and thus able to set their strategic course in the business environment of Germany (Fig. 5.2).

At the same time, the survey shows that more than 82% of the remaining 29% are in the hands of German shareholders and only 7% in the hands of non-European interested parties (Fig. 5.3).

With regard to shareholder companies, in about 15% of the cases venture capital investors are involved. Consequently, these holdings have to be taken into account not primarily for company-specific, business-strategic reasons, but for purely financial reasons.

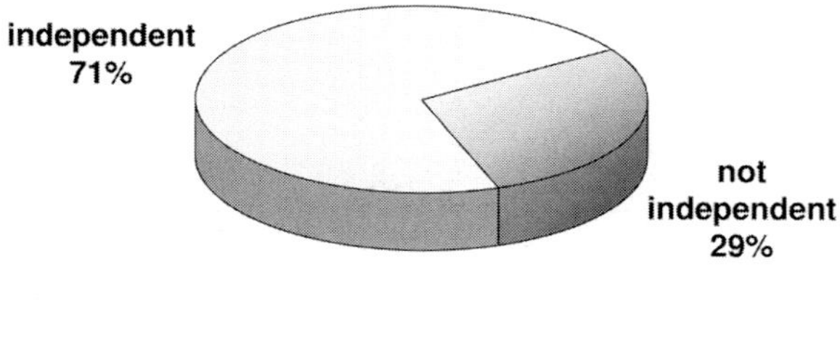

Figure 5.2 Economic independence of the companies polled.

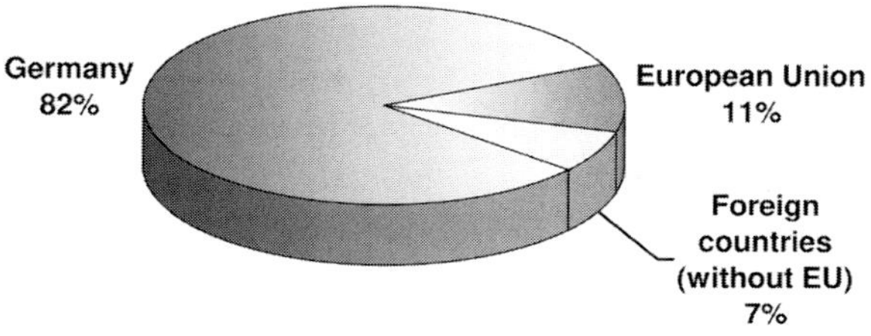

Figure 5.3 Location of the shareholders.

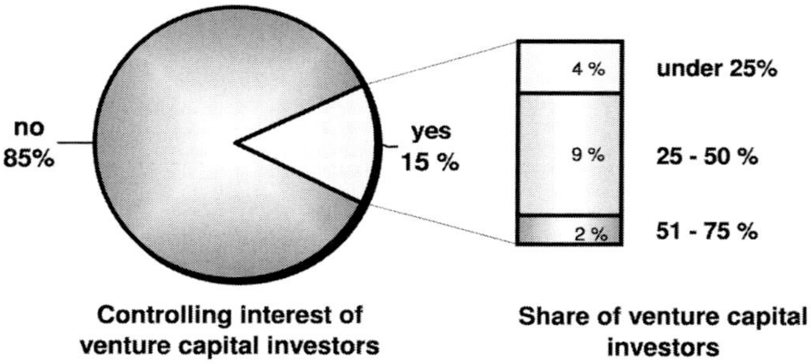

Figure 5.4 Share of venture capital investors.

Accordingly, Fig. 5.4 illustrates that at 2%, the controlling interest of more than 50% is the absolute exception. The majority of the equity interests range between 25 and 50%.

Business activity

Apart from the economic position of the companies, even the priorities regarding the business activities are an indicator for the companies' orientation. Here it is clear in which fields nanotechnology is already regarded as a relevant technology. This is based on the hypothesis that the majority of the companies researches and develops in the fields of their traditional markets (Fig. 5.5).

Analyzing the business segments shows that companies active in nanotechnology have their markets mainly in the chemical industry and in the manufacturing of measuring, control and navigation instruments. This absolutely confirms the results of other surveys.

Distribution of customers

Nanotechnology as a relatively young field of technology is regarded as a future technology both in Germany and in Europe, as well as in Asia and the USA. Against this backdrop, the question was posed concerning the position of the primary customers of the company within a global context (Fig. 5.6).

All in all, a large part of the total turnover[1] of the companies in question is made in Germany, followed by Europe (excluding Ger-

1) Turnover inclusive and exclusive of nanotechnological products.

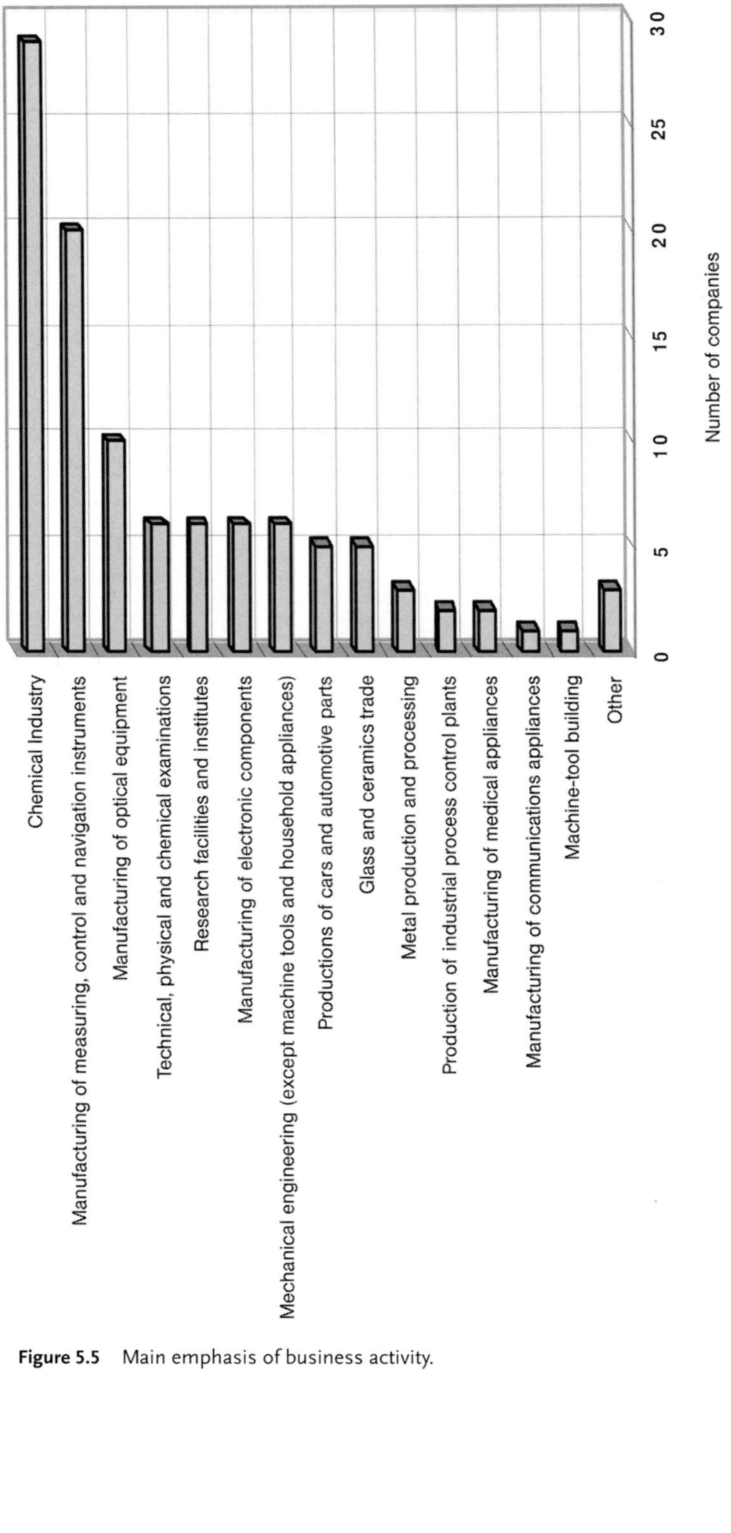

Figure 5.5 Main emphasis of business activity.

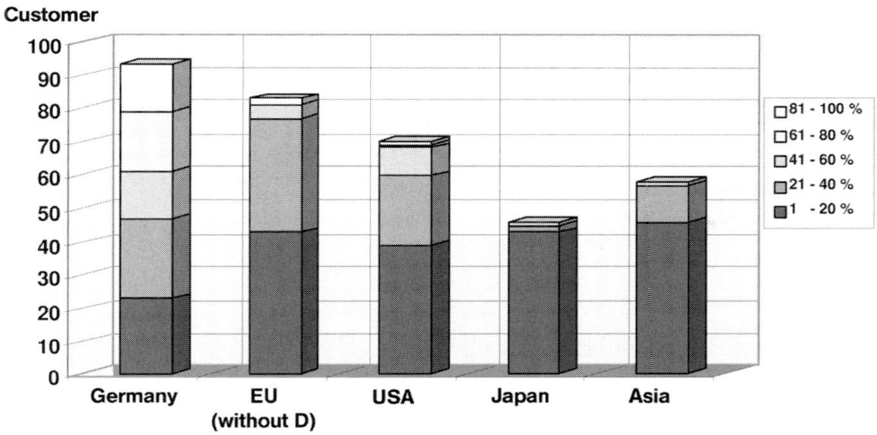

Figure 5.6 Distribution of customers of the companies polled.

many) and the USA. Asia and especially Japan play a secondary role for the companies in the sample.

Important information is provided by the connection between the location of the primary customers and the priorities of business activities. It is apparent that in the fields of metal production and processing, machine-tool building, in the manufacturing of medical appliances, measuring, control and navigation instruments and in technical, physical and chemical examinations, more than 50% of the customers are located in Germany.

In the business segment of chemical industry, glass and ceramics trade, mechanical engineering (excluding machine tools and household appliances), the share of German customers lies at about 50%. The rest of the business segments have higher shares of foreign customers Here, Asian customers are of great importance, particularly in the fields of production of industrial process control plants (50%), mechanical engineering and the manufacturing of medical appliances (22% each) as well as with regard to the manufacturing of electronic components (20%).

In contrast to this, the strongest contacts of customers with the USA are found above all in the business segments of manufacturing of industrial production plants and electronic components (26% each) as well as in the chemical industry (25%).

A closing description of the companies contained in the sample can be obtained from the distribution of the turnover and the number of employees.

Business turnover and business start-ups

The business turnover (Fig. 5.7) gives already rise to the suspicion that the number of business start-ups may be high. About 32% of the companies in the sample show a turnover of up to 2 million EUR in

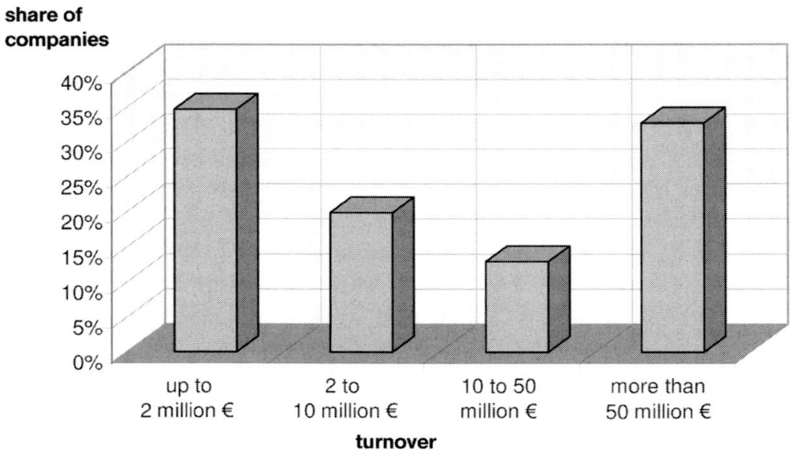

Figure 5.7 Business turnover in the financial year 2001

2001. Nearly 20% of the other companies range between 2 and 10 million EUR. However, about 30% of the companies in the sample declare a turnover of more than 50 million EUR.

SME Even more distinct is the share of the SME in the sample when the number of employees is analyzed. About 35% of the companies have up to 20 employees. For altogether almost 68% of the companies participating in the survey, the EU definition for a SME[2)] with up to 250 employees is still applicable (Fig. 5.8).

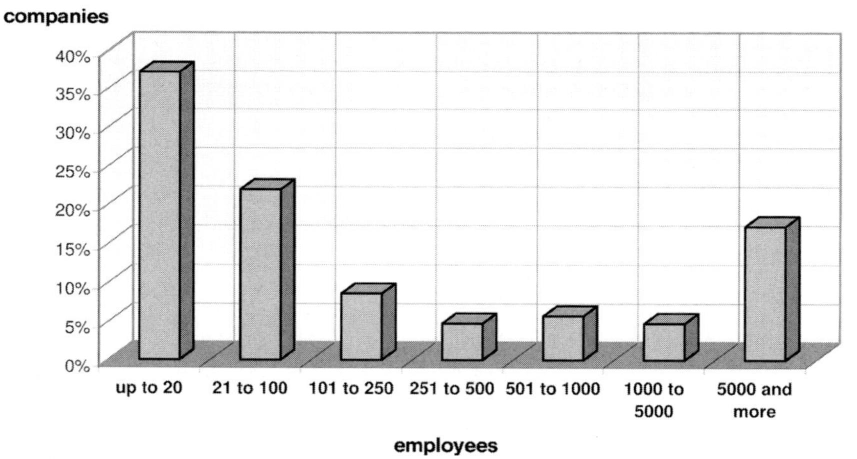

Figure 5.8 Number of employees per company.

2) According to the definition of the EU, SME are companies with up to 250 employees and an annual turnover of up to 50 million EUR and 42 million EUR balance sheet sum.

The data acquisition in the sample was mainly obtained from the total company (75%). Only a small number answered the questions with regard to a company division (10%) or a staff unit (15%).

5.4 The Current Importance of Nanotechnology

In the survey, the questions were investigated as to since when and in which way companies are dealing with nanotechnology. Here the most frequent starting time for nanotechnology turned out to be the period from 1996 to 2000. In this period, both observation of the nanotechnological scene and the companies' own R&D, as well as the application of nanotechnology in products, experienced their sharpest increase. Discussions concerning market-relevant relations (1996) initiated by the BMBF, for example, and the establishment of the publically-funded competence centers (1998) for the nanotechnological sector fall into this period as well.

Growing importance since the middle of the 1990s

Considering the developments of the different activities over the years, this picture becomes even clearer. The analysis of the activities shows plainly their shift from observation to realization over this period of time (Fig. 5.9).

This tendency is also confirmed by the development of the turnover since 1996. Comparing the turnover of products with nanotechnological input made between 1996 and 2001, it is noticeable that the turnover in 2001 exceeded that of 1996 in all turnover classes. Moreover, the analysis shows clearly that both the number of companies re-

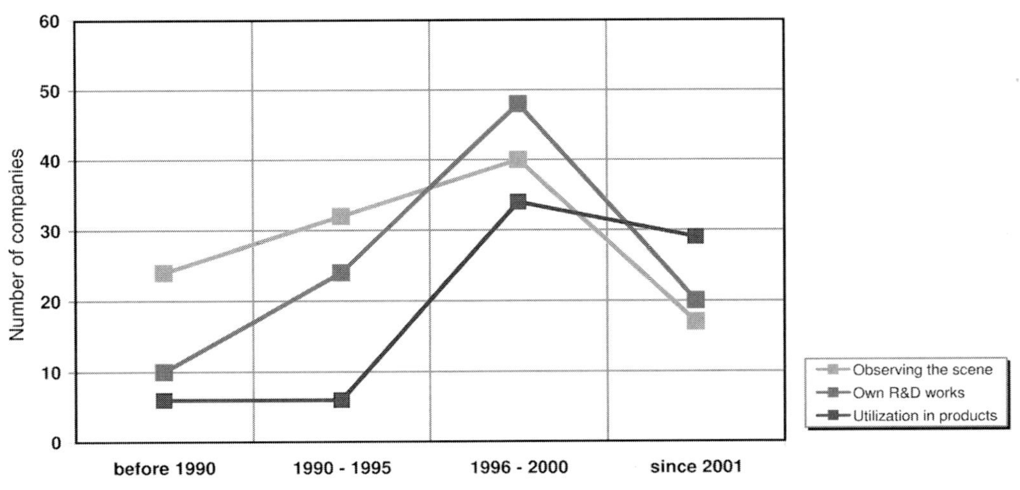

Figure 5.9 Attitude to nanotechnology.

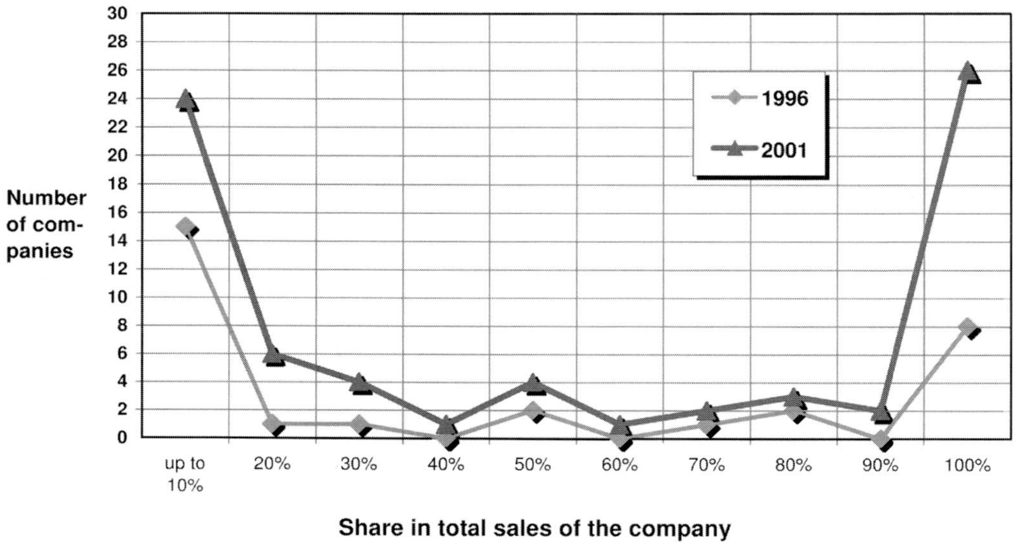

Figure 5.10 Nanotechnology-related sales share.

alizing first turnovers with products with nanotechnological input and the number of companies making their total turnover with such products has grown considerably (Fig. 5.10).

For the evaluation of the sales figures it is reasonable to subdivide them into two categories:

- turnover from the manufacture of nanotechnological products;
- turnover from products that nanotechnology is used in.

Number of active market players increased

Here we should emphasize that 66% of the companies describe themselves as manufacturers and about 29% as purely users of nanotechnological products[3]. This allows a first conclusion about the diffusion of nanotechnological knowledge. Consequently, the group of purely users of nanotechnological primary products has played a less important part so far compared to the active manufacturers of such components. This pattern is typical for a diffusion process. Therefore it is a good sign for ongoing diffusion that the number of users who are purely market observers, who achieve no turnovers with nanotechnological primary products, has decreased from 71% in 1996 to 25% in 2001. Hence, the number of active market players has increased. This is also verified by the correlation calculation of sales share and primary products.

3) 5 percent did not make any statement.

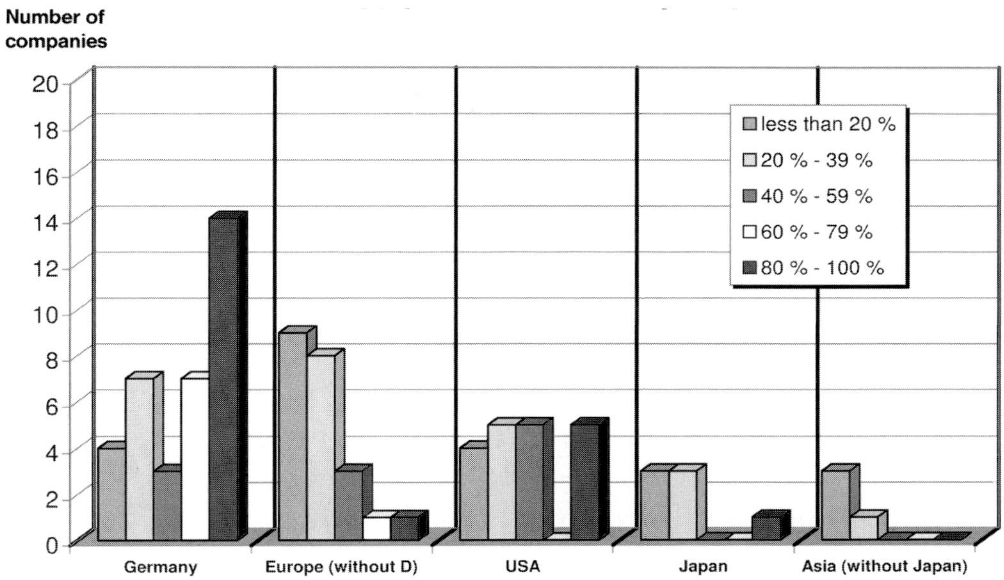

Figure 5.11 Supply of nanotechnological products.

Analysis of the economic areas that the respective nanotechnological primary products are obtained from, reveals that the German economic area is still the largest supplier market, followed by Europe (excluding Germany) and the USA (Fig. 5.11).

A description of the priorities in the business areas of the companies was given in Section 5.3. An interesting aspect is the correlation of these priorities with the priorities in the activities of companies where nanotechnology plays a central role nowadays. The question regarding the relevance of application fields was answered as shown in Fig. 5.12.

Here again, chemistry, materials and process engineering are at the top, followed by medical technology/health and I&C. This can be regarded as proof of the fact that companies dealing with nanotechnology are acting in such fields that represent their core competence even without nanotechnological components.

International competition

This is reinforced when it comes to the issue of competitors in the application fields. Here it is obvious that in the field of chemistry the main competitors are presumed to be in Germany and in the USA. In contrast to this, players in medical technology/health, in I&C and in measurement technology see their competition above all in the USA (Figs. 5.13–5.17).

We may begin to draw conclusions about the development of diffusion in Japan, the USA and Europe, where a typical pattern seems to

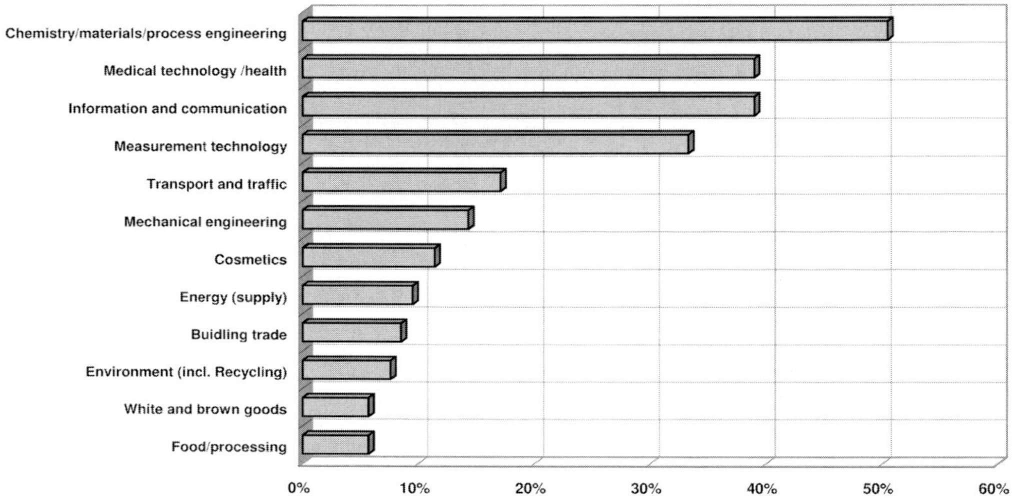

Figure 5.12 Characteristics of today's fields of application.

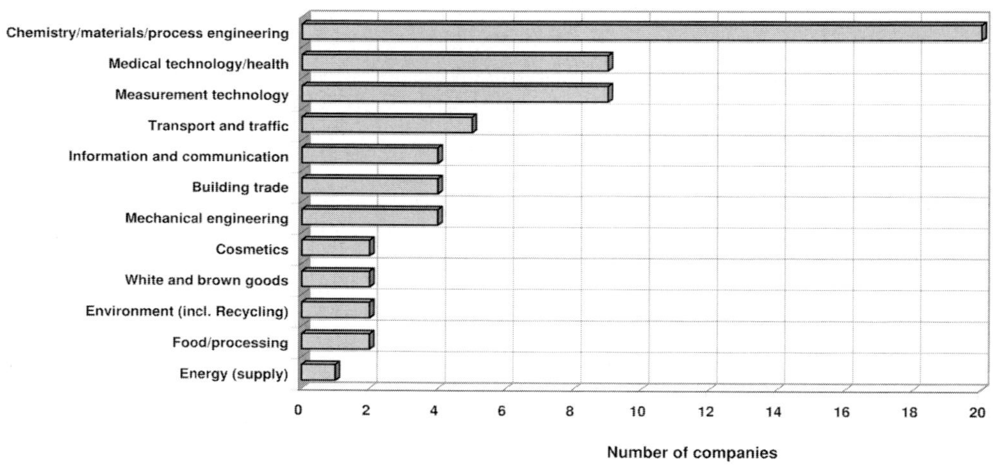

Figure 5.13 Number of German competitors in the field of application.

reappear. While the analysis of the relevance of primary products suggests that Germany has reached a good position in the field of research, companies in the USA seem to react faster when it comes to the conversion of research results into products (e.g. in the field of I&C). Just as interesting is the fact that currently neither Japan nor the rest of Asia are really regarded as competitors in the diffusion of nanotechnology.

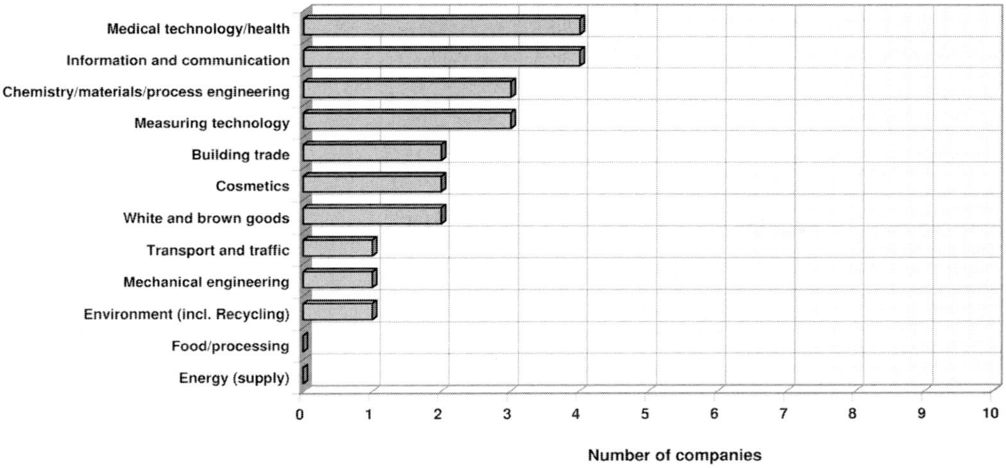

Figure 5.14 Number of European competitors (except Germany) in the field of application.

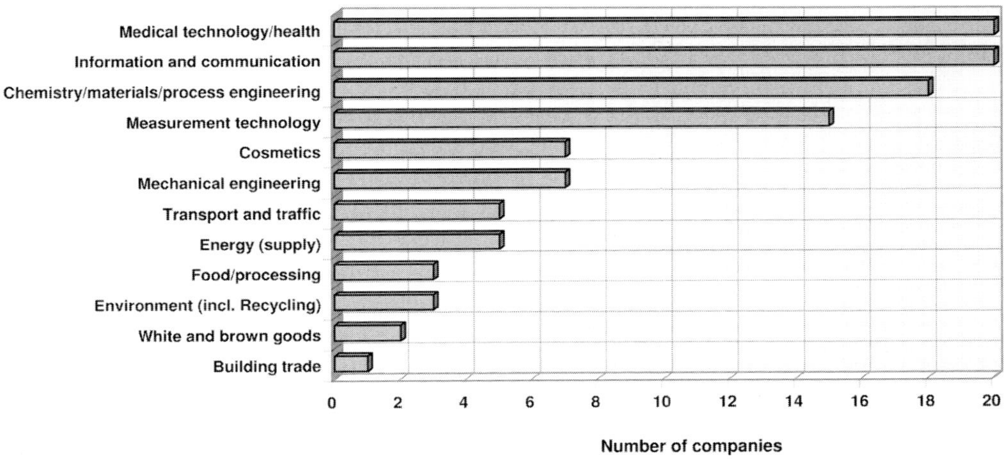

Figure 5.15 Number of American competitors in the field of application.

International comparison of research and realization

This result is partly confirmed by analysis of the assessment of the power of potential competitors, R&D and the realization in the market (Figs. 5.18 and 5.19). Here it can be seen that especially in R&D, Asia (excluding Japan), Japan and Europe (excluding Germany) lag behind the USA and Germany. However, the persons questioned believed that the USA and Japan are better able than Germany to convert nanotechnology into products.

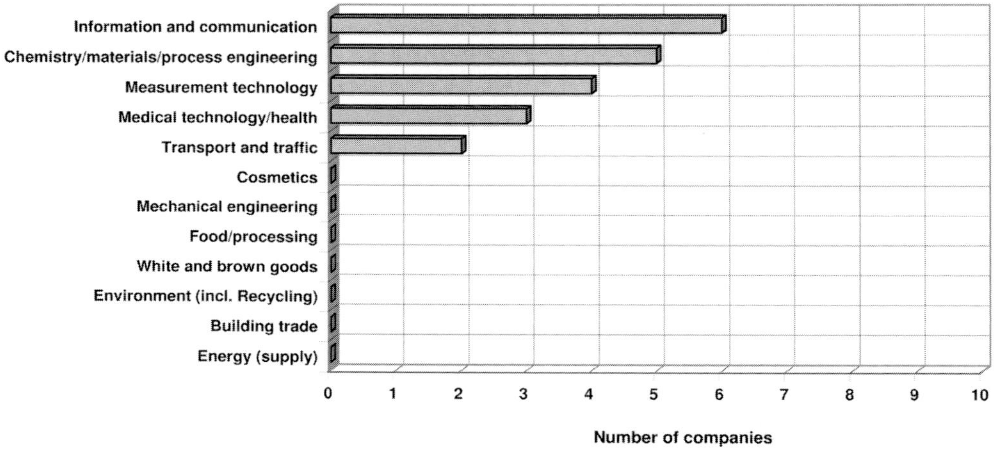

Figure 5.16 Number of Japanese competitors in the field of application.

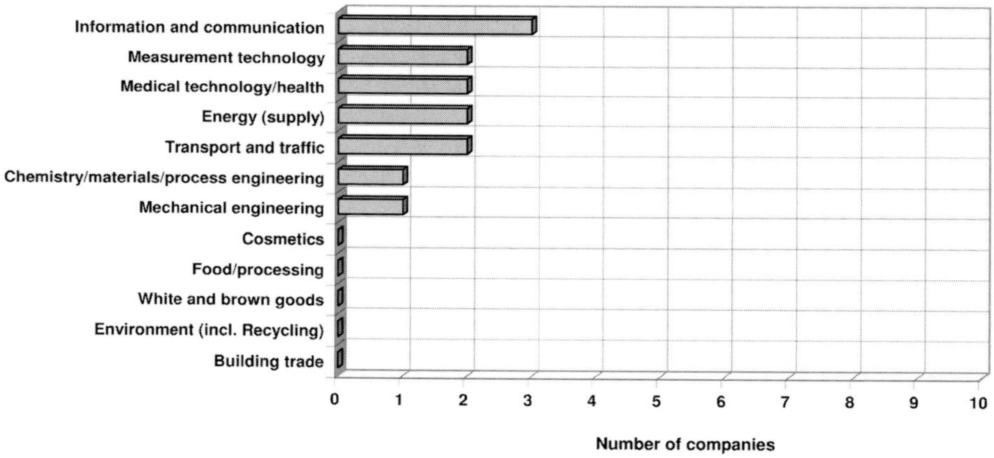

Figure 5.17 Number of Asian competitors (except Japan) in the field of application.

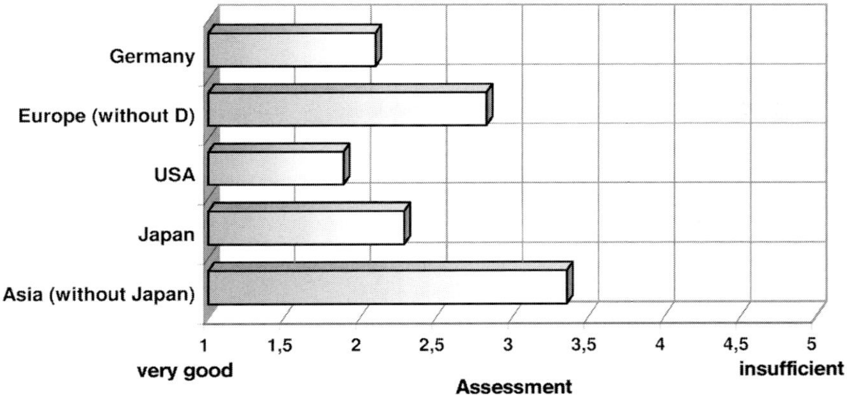

Figure 5.18 International comparison of the perception of research power.

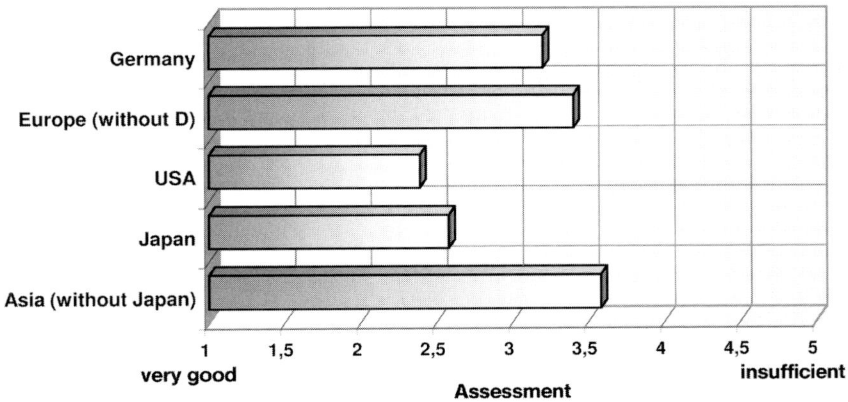

Figure 5.19 International comparison of the perception of marketing power.

This is found again in the analysis of the strongest competitors (Fig. 5.20).

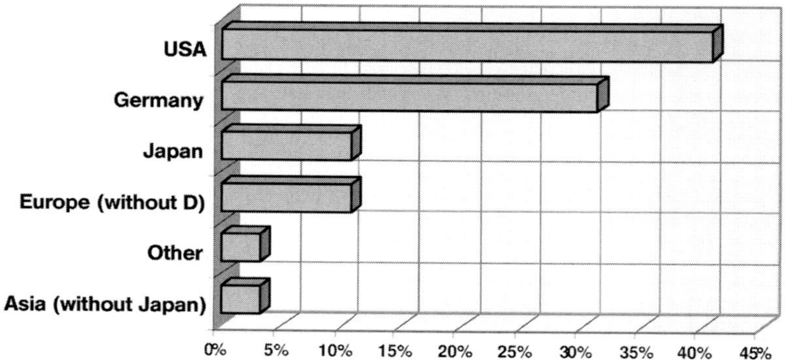

Figure 5.20 Location of headquarters of the strongest competitors.

Nanotechnology no technological "playground"

The importance German companies attach to nanotechnology has been analyzed. It is clear that the companies in the sample see above all the economic opportunities of nanotechnology and do not regard it as a technological "playground". More than 75% of the companies believe in their ability to conquer new markets with nanotechnology. More than 60% of the companies consider nanotechnology to be a crucial competition factor, and a chance to improve their technological competitiveness (Figs. 5.21–5.23).

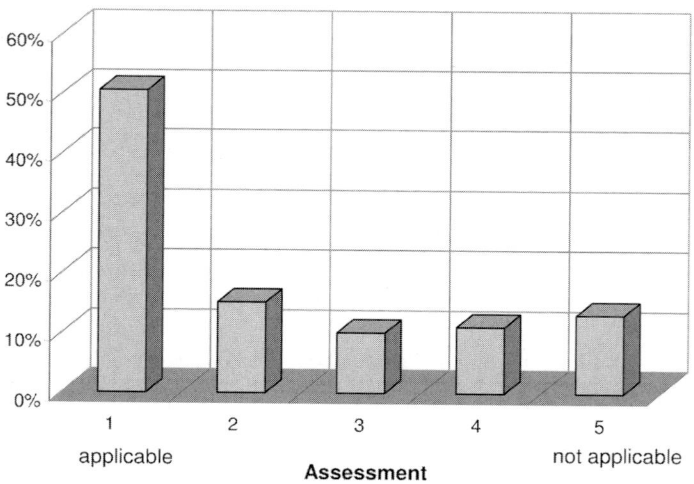

Figure 5.21 Evaluation of nanotechnological know-how as a decisive competitive factor of the company.

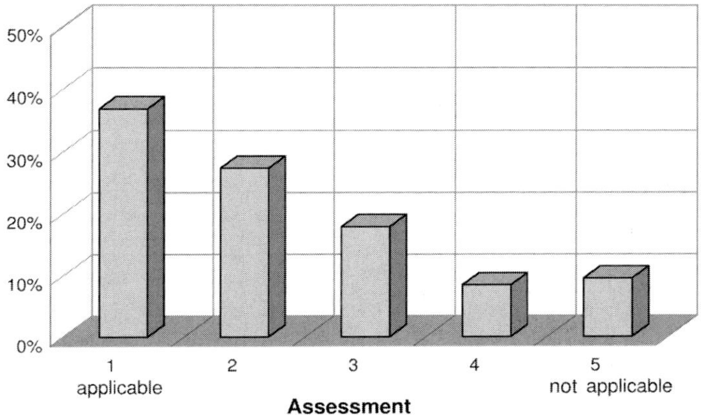

Figure 5.22 Evaluation of nanotechnological relevance with regard to the company's competitiveness in traditional markets.

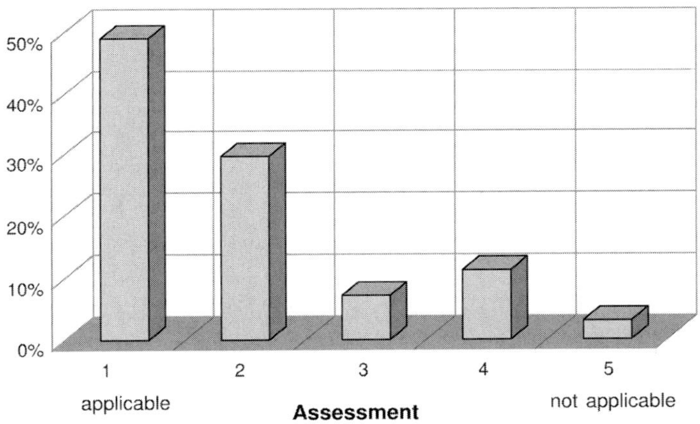

Figure 5.23 Evaluation of nanotechnological relevance with regard to the company's opening-up of new markets.

In contrast to this, the companies disagree with the statement that nanotechnology would offer a new field for experiments, just as they reject in a more moderate way the statement that it would extend technological competence (Fig. 5.24).

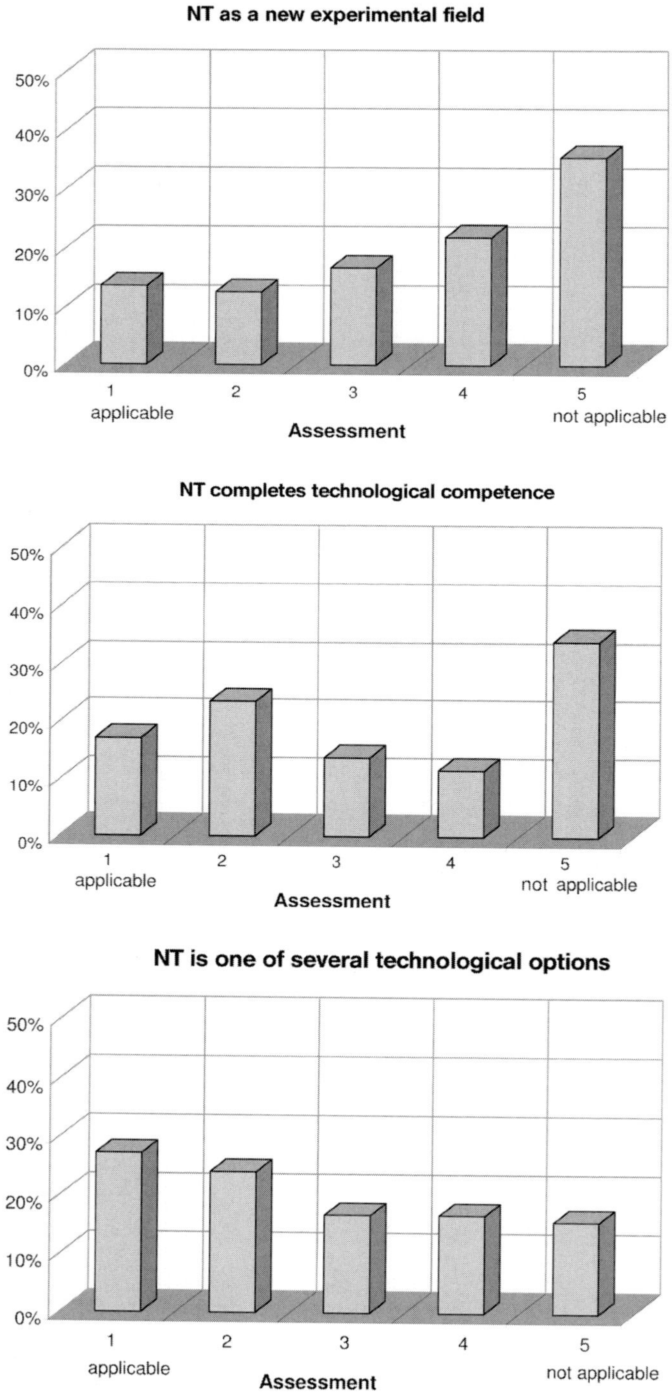

Figure 5.24 Evaluation of nanotechnological relevance with regard to the technological competence of the company and the attitude towards its role in the company.

5.5 Future Developments in Nanotechnology

Most of the companies regard the prospects of nanotechnology as positive. Almost 90% of the companies will enhance their activities in nanotechnology, 30% of them even considerably (Fig. 5.25).

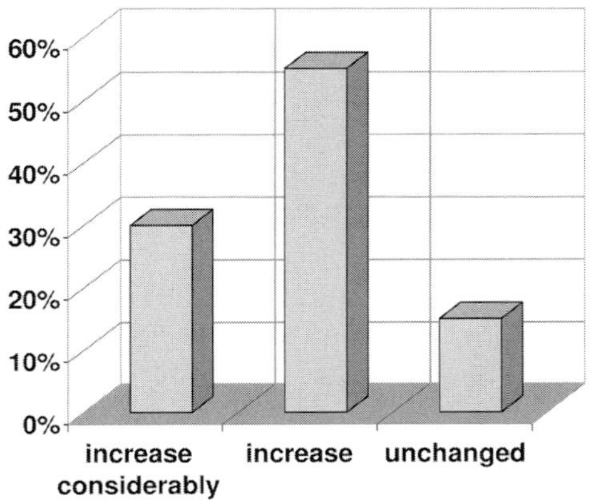

Figure 5.25 Development of nanotechnological activities.

Growing manpower requirements

This is accompanied by an increase in employment. Only 18% of the companies polled do not see growing manpower requirements for their nanotechnological activities. The rest of the companies do expect a growing requirement of manpower (Fig. 5.26).

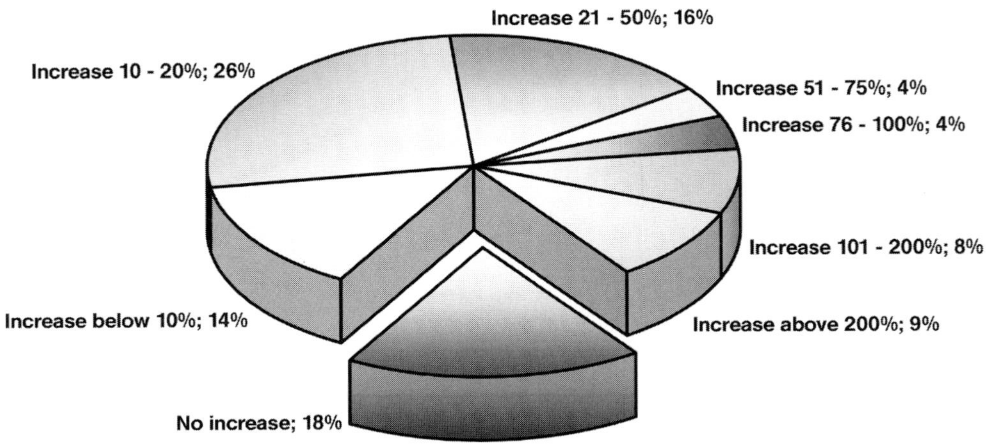

Figure 5.26 Increase in labor employment in nanotechnology in the company up to 2006.

Table 5.1 Correlation between size of turnover (2001) and expected additions to workforce (2006).

Which turnover did your company obtain in the financial year 2001?	Additions to workforce: Do you expect an increase in your labor employment in the field of nanotechnology by 2006?		Total
	yes	no	
up to 2 million EUR	28	7	35
2 to 10 million EUR	16	4	20
10 to 50 million EUR	8	3	11
more than 50 million EUR	26	5	31
Total	78	19	97

Here it is significant that it is the small and large-scale companies that more strongly predict an increase in staff required. Table 5.1 reflects the results for this relationship.

When searching for future markets of nanotechnology it was not possible to consider specific products, as the companies could hardly have predicted as far ahead as 2006. Consequently, the survey focused on the examination of the fields of application relevant in the year 2006. To obtain a ranking of the answers, the persons polled were asked to set three priorities in the relevance of future application fields (Fig. 5.27).

The ranking shows that most future relevance is attributed to the application fields ICT, chemistry/materials/process engineering, and medicine and life sciences, regarding both the first and the further priorities. Obviously, companies see the best market potential in these fields of application by the year 2006.

Application fields with promising future

In order to achieve an assignment of the application fields with a promising future, and the functional elements and technologies necessary for their realization, corresponding links were called up. The interviewees were about the connection between the three fields of application that they regarded as the most important ones, and the required functions and technologies. With this, it is possible to learn about the functional requirements the nanotechnological applications have to fulfill in the respective fields of application. On the other hand, with the further link to the required technologies, it is possible to show which technological orientations of the companies are relevant for the realization of their ideas in the respective application fields.

When choosing the possible functions, attention was paid to polling as comprehensive a spectrum as possible. Altogether 19 functions were offered according to their relevance for the application fields.

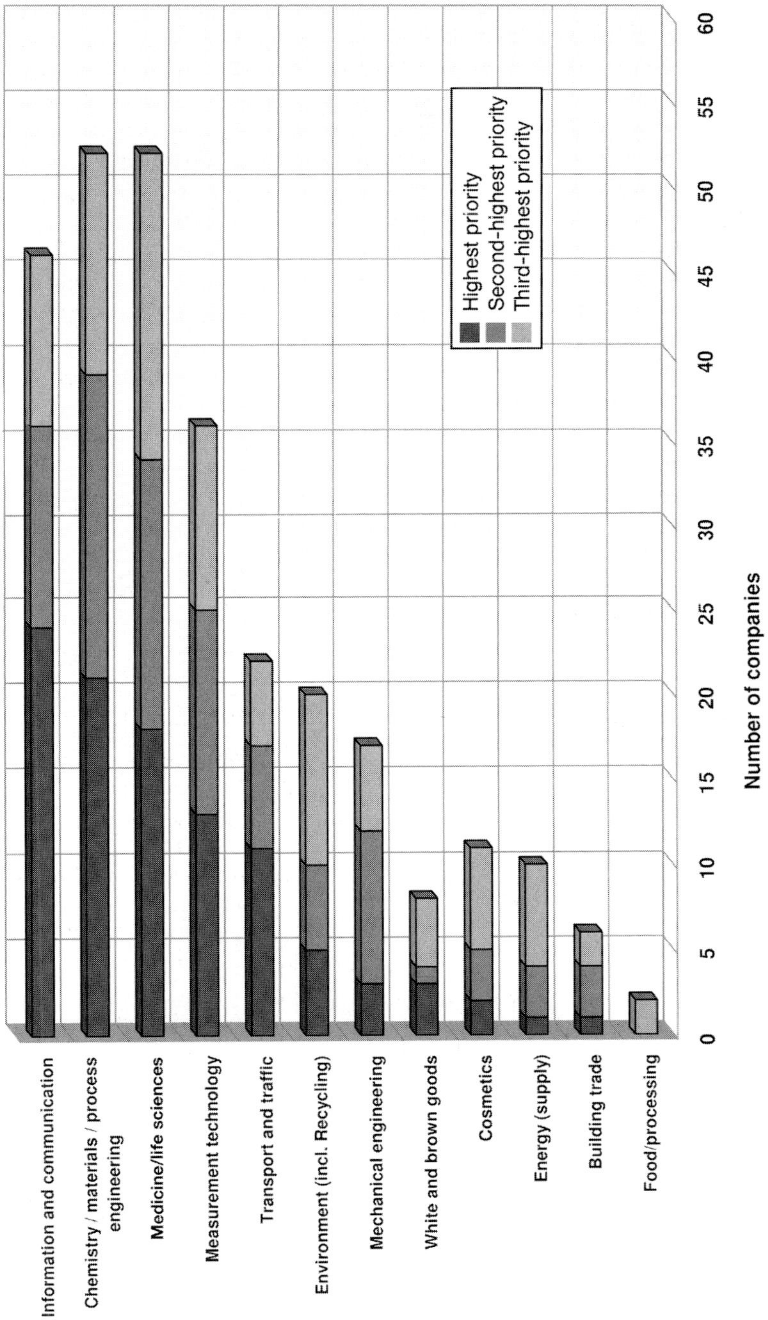

Figure 5.27 Future relevance of application fields.

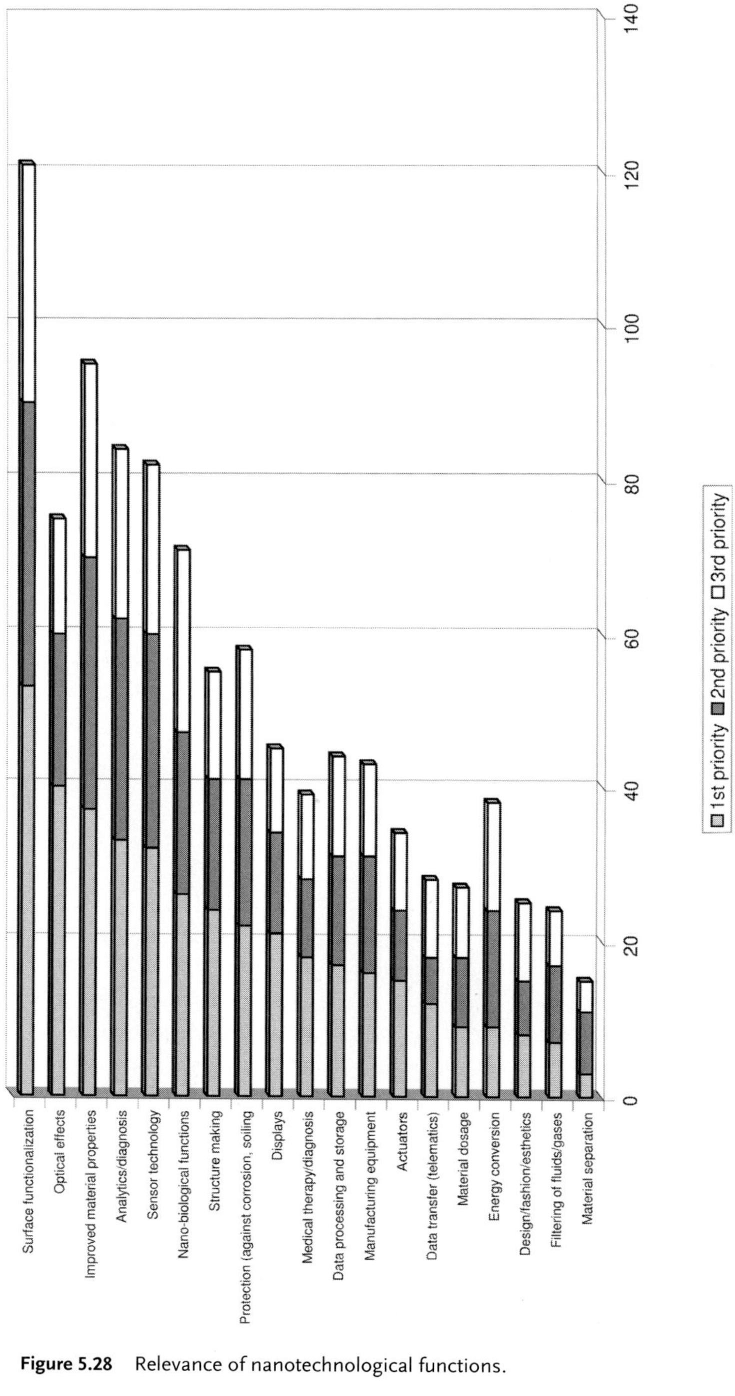

Figure 5.28 Relevance of nanotechnological functions.

Moreover, further functions could be added in an open question, an opportunity taken up by 12 interviewees (Fig. 5.28).

The analysis of the distribution by functions shows a strong orientation towards the following functions: surface functionalization, optical effects, improved material properties, analytics/diagnosis, sensor technology and nanobiological functions. Examining the frequencies with regard to the three priorities, then it is only the order of the given functions that shifts.

Concerning the required technologies, there were altogether 16 possible choices. Here again it was possible to give an open answer, which was done by 29 interviewees (Fig. 5.29).

The analysis of the frequencies by priorities makes clear that the technologies of thin film deposition methods, microscopy, self-assembly and optical lithography show the highest frequency density. However, all-in-all, the field in the ranking is very dense so that the frequencies alone only permit very limited conclusions.

A more informative value is provided by the relationships between the variables. Here, first the attributes “function“ and “fields of application“ were analyzed, then the attributes “fields of application“ and “technologies“. Table 5.2 reflects the correlation between fields of application and functions. In this case the analysis by priorities was disregarded, as then the frequency density would be too low, which would have negative effects on the informative value of the statements.

We begin to see indications regarding the question of which functional requirements in the respective fields of application are actually needed. So the central functions in the most densely frequented application field of ICT are found in the segments data processing and storage, displays and optical effects. This can be explained by the growing demand on the performance of IT hardware and software as well as with the growing demand for high-resolution user interfaces that are also light and robust. (Fig. 5.30).

ICT

In the application field “chemistry/materials and process engineering” the central functional properties are improved material properties and surface functionalization, followed by protection function and optical effects. From this, first indications can again be derived (Fig. 5.31).

Chemistry, materials and process engineering

In medical technology, the central functions are seen in medical therapy/diagnosis, the utilization of nanobiological properties and analysis/diagnosis. This can also be explained by the application of nanotechnological products in the health sector. Obviously, the interviewees expect the highest yields from products with nanotechnological elements in the diagnostic field of application (Fig. 5.32).

Medicine and life sciences

It is possible to make connections between application fields and nanotechnologies in the same way as to relations between functions

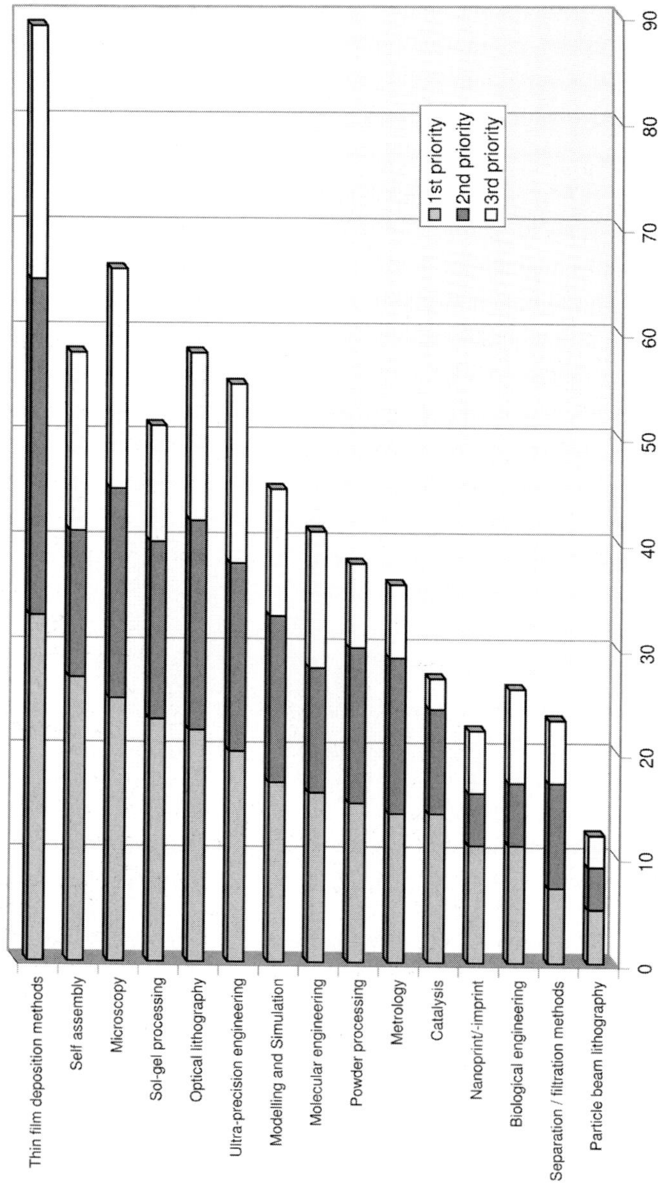

Figure 5.29 Relevance of different nanotechnological attributes.

Table 5.2 Fields of application in priority 1–3 in connection with the functions in priority 1–3.

	Information/ Communication	Medicine/ life sciences	Cosmetics	Transport and traffic	Chemistry	Environment	Energy	Mechanical engineering	Measurement technology	Building trade	Food	White/brown goods	Total
Functions													
Analytics/diagnosis	7	28	4	1	15	1	0	0	19	1	2	0	**78**
Medical therapy/diagnosis	0	34	1	0	1	0	0	0	1	0	0	0	**37**
Surface functionalization	16	23	3	11	33	3	4	5	7	4	1	1	**111**
Displays	21	3	1	3	4	0	1	0	6	0	0	2	**41**
Energy conversion	5	0	1	4	7	3	9	3	2	1	0	1	**36**
Production equipment	11	6	0	3	6	1	3	5	4	1	0	0	**40**
Exploitation of nanobiological properties	2	31	5	1	14	2	2	1	3	2	2	1	**66**
Date processing and storage	22	3	1	2	4	0	1	1	6	0	1	0	**41**
Data transmission	12	3	1	4	3	0	1	0	3	0	1	1	**29**
Material separation	1	4	1	0	7	0	0	1	0	0	0	1	**15**
Sensor technology	12	16	2	7	15	4	3	3	16	0	0	3	**81**
Actuators	7	6	0	4	5	1	1	2	7	1	0	0	**34**
Material dosage	1	7	4	1	6	0	0	2	2	1	0	1	**25**
Optical effects	20	3	5	8	18	0	1	1	13	1	0	2	**72**
Filtering of fluids and gases	2	7	2	1	7	1	1	0	0	0	0	1	**22**
Protection	4	2	3	9	21	1	0	5	2	5	0	0	**52**
Improved material properties	10	7	2	13	34	4	4	8	5	4	1	1	**93**
Structure generation	15	4	0	4	12	0	3	6	3	2	1	0	**50**
Design/fashion/esthetics	4	1	5	3	7	0	0	0	1	1	0	1	**23**
Others	1	3	0	1	2	0	0	0	1	0	0	1	**9**

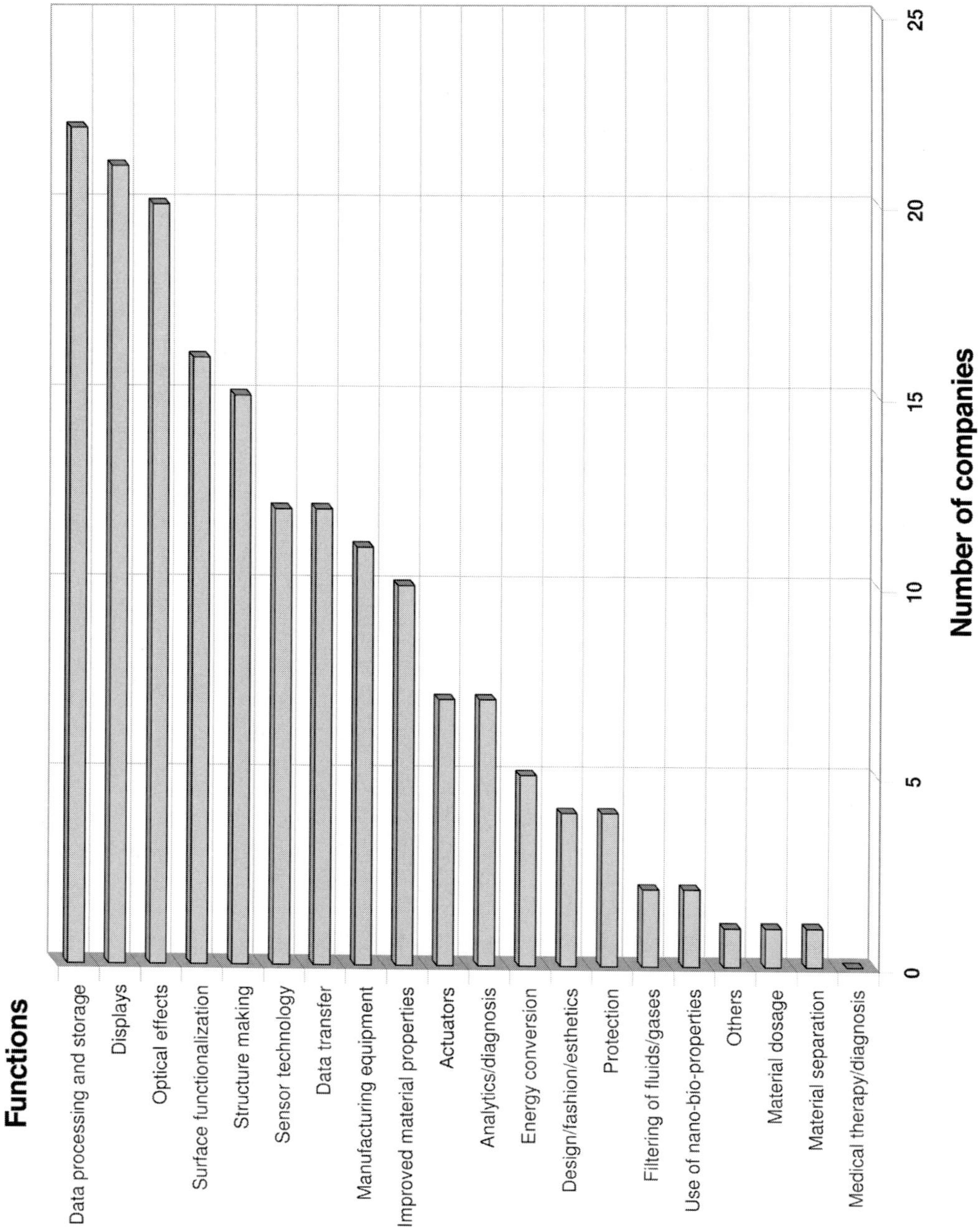

Figure 5.30 Importance of different functions in the application field "Information and Communication".

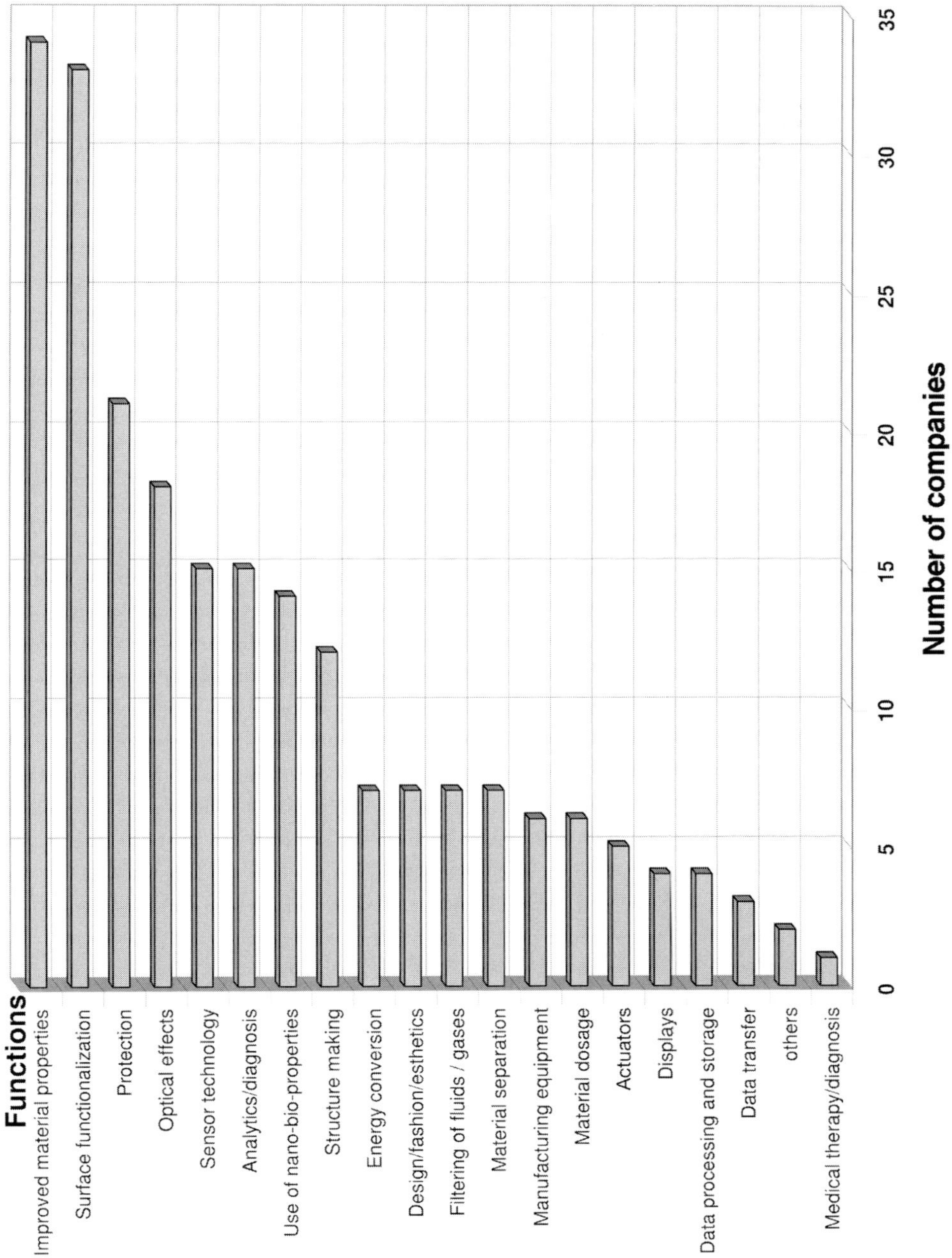

Figure 5.31 Importance of different functions in the application field Chemistry/materials and process engineering.

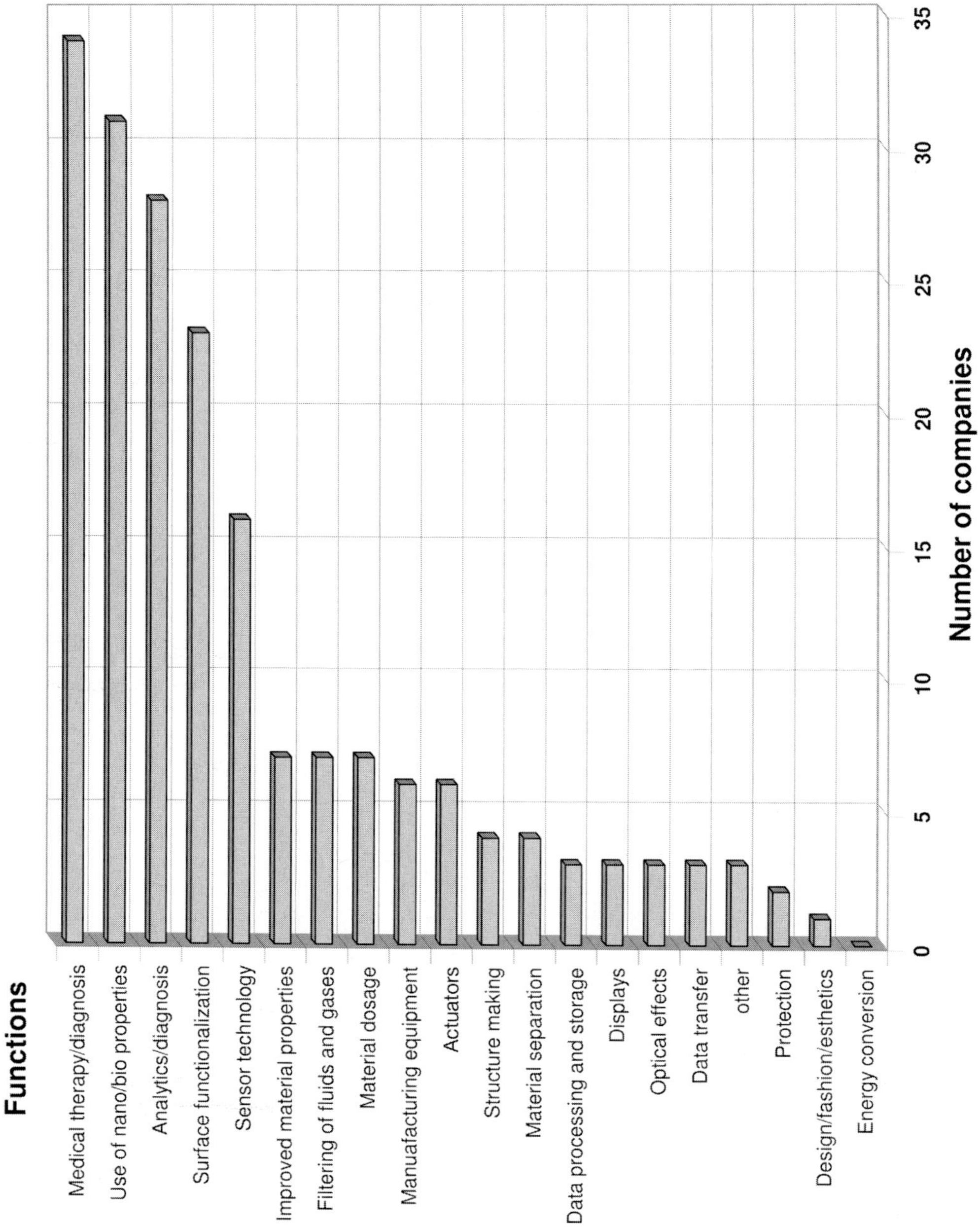

Figure 5.32 Importance of different functions in the application field Medical technology/health.

and application fields. Here, too, analysis is based on the frequency density of the respective attributes between the different technologies and application fields that were called up.

The forms of utilization given in Table 5.3 related to fields of application and technologies plainly show the range of application of the individual technologies and with this their likely prospects. It is obvious that the utilization range in medical technology, chemistry/materials and process engineering, information and communication technology as well as in transport and traffic is particularly well developed. In contrast to this, the prospects of the technologies in the food industry are the most limited.

In detail, especially convincing results are achieved in the field of medicine and life sciences, which crystallize clear prospects of the technologies microscopy, biological engineering, thin film deposition methods and self-assembly (Fig. 5.33).

Table 5.3 Application fields in priority 1–3 in connection with the technologies in priority 1–3.

Technologies	Information and Communication	Medical technology	Cosmetics	Transport	Chemistry	Environment	Energy	Mechanical engineering	Measurement technology	Building trade	Food	White/brown goods	Total
Separation/filtration methods	1	4	1	3	5	2	3	0	2	0	0	1	**22**
Biological engineering	1	13	0	0	6	0	0	0	2	0	1	0	**23**
Sol-gel processing	3	7	2	10	14	1	0	2	1	4	0	0	**44**
Powder processing	4	3	3	6	12	1	2	4	0	1	0	0	**36**
Particle beam lithography	6	0	0	0	1	0	1	0	3	0	0	0	**11**
Catalysis	6	2	1	6	7	1	2	0	0	1	0	1	**27**
Self assembly	7	12	2	5	13	2	2	2	3	2	2	1	**53**
Nanoprint/-imprint	8	3	0	2	0	0	0	2	2	0	0	1	**18**
Molecular engineering	8	11	0	3	8	1	3	0	1	0	2	1	**38**
Metrology	9	4	0	1	3	1	0	2	13	1	1	0	**35**
Microscopy	11	14	2	3	15	1	0	5	9	1	1	1	**63**
Modeling and simulation	12	6	1	6	7	0	2	2	4	0	0	2	**42**
Ultra-precision engineering	17	11	0	2	5	0	2	6	8	0	0	0	**51**
Optical lithography	20	10	1	5	2	1	2	2	10	0	0	2	**55**
Thin film deposition methods	21	13	3	8	14	5	6	2	7	2	0	3	**84**
Others	3	7	2	2	6	2	1	0	2	0	0	0	**25**

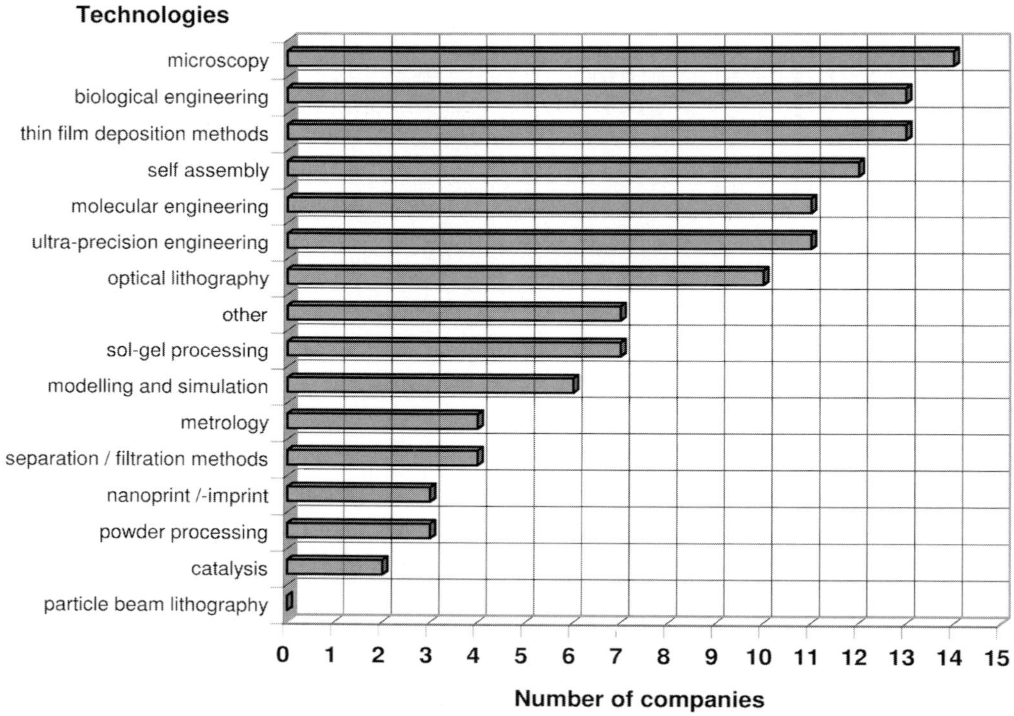

Figure 5.33 Importance of different technologies in the application field of medicine and life sciences.

Here are indications of future fields that, on the one hand, might be seen in new methods of microscopy, the market volume of which should certainly not be overestimated, and on the other hand in the fields of self-stimulating biological systems that might mainly be found in the skin-related field.

Future fields

In chemistry/materials and process engineering, again microscopy together with sol-gel-processing and the thin film deposition method are the most distinctive fields. But also powder processing and self-assembly have a high potential for this field of application (Fig. 5.34).

All-in-all it is clear that, apart from medical technology, even in chemistry/materials and process engineering the range of the technologies pursued is extremely well developed. Lithographic processes are still pursued with a relative frequency in this field.

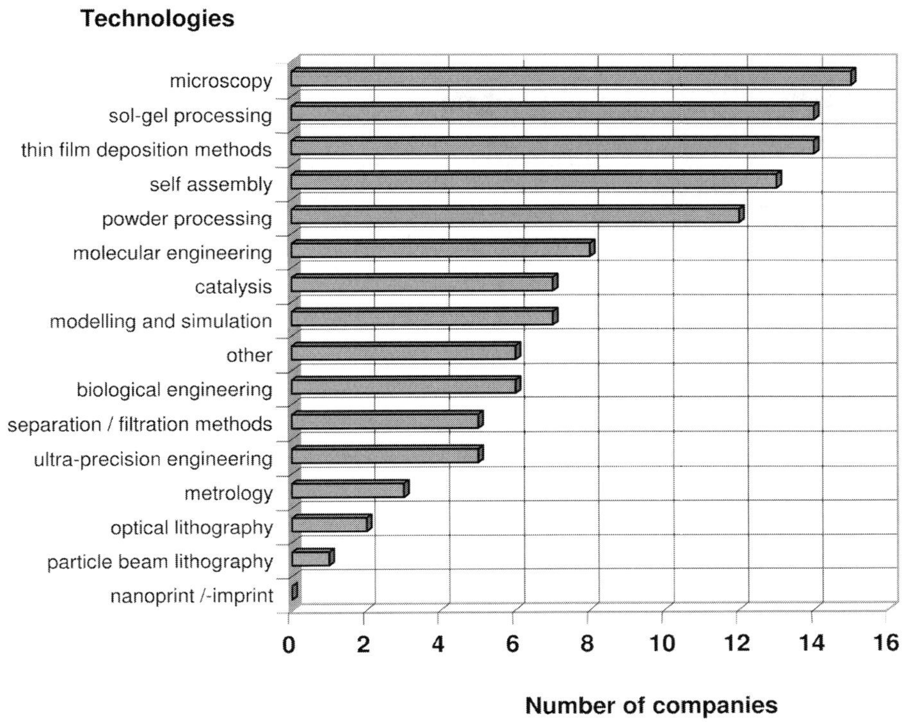

Figure 5.34 Importance of different technologies in the application field chemistry/materials and process engineering.

5.6 Future Steps and Obstacles for the Development of Nanotechnology

5.6.1 Next Steps to be taken in Nanotechnology

Current phase of work in the company

In the following, the results regarding the companies' measures planned for the future development of nanotechnology are presented as well as the obstacles they are faced with. The results refer to the respective field of application the interviewee attached the highest priority to. It was necessary to focus on one field to obtain unambiguous answers.

The problematic nature of a close relationshipbetween R&D and market placement is made clearer by the phase the company is currently going through. We see that in the broad field of nanotechnology, there are companies still dealing with R&D while others have already placed products on the market (Fig. 5.35).

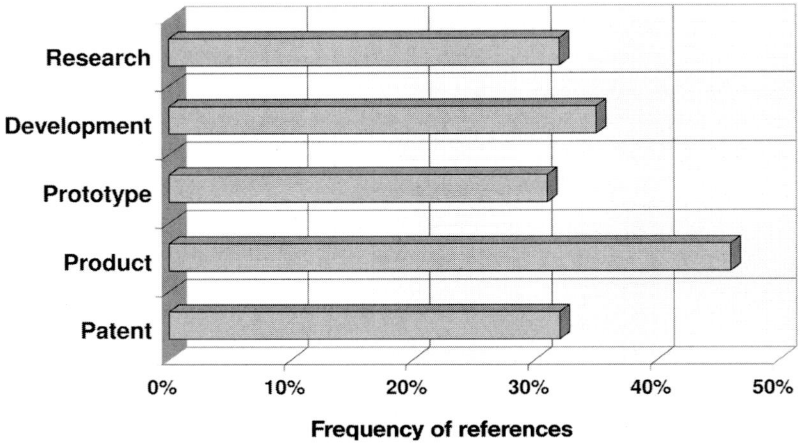

Figure 5.35 Current process phase in the development of nanotechnology of the companies polled (multiple answers).

The values for the different phases in the technology and innovation cycle are therefore commensurately high. This result would actually be very pleasant if the sale of the products or at least the measures necessary for that, together with the market placement, were initiated, too. But the analysis of the question about the first steps initiated shows, however, that all the measures enabling the quick market placement, such as the acquisition of property rights and the establishment of an international distribution network, have only low frequency densities (Fig. 5 36).

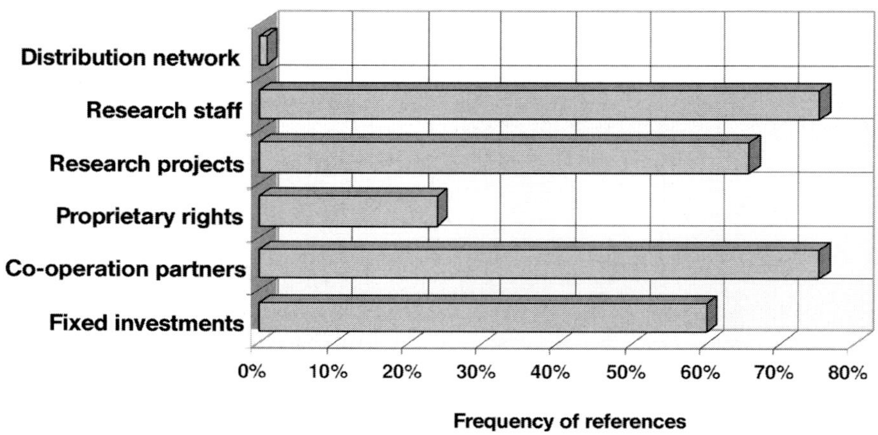

Figure 5.36 Steps for the realization of nanotechnological projects in the future (multiple answers).

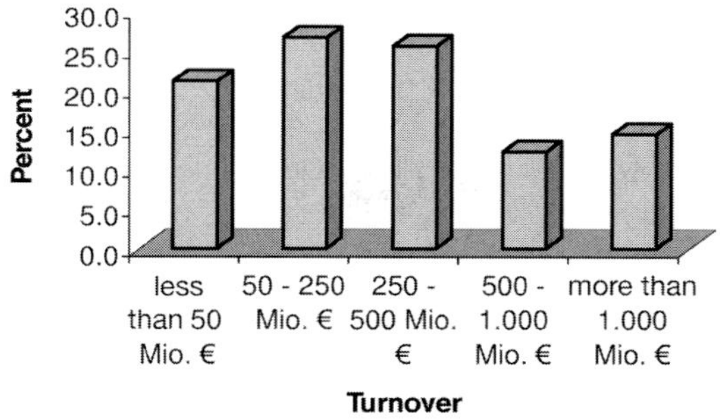

Figure 5.37 Expected market volumes for specific products in the year 2006.

The strong emphasis on the provision of research staff, the establishment of contacts to cooperative partners and real investment transactions lead to the conclusion that currently, the main concentration is still on the research effort which, indeed, corresponds to the time horizon up to 2006 (Fig. 5.37).

Mainly research efforts

The companies polled have an absolutely positive attitude towards the potential of their core products in nanotechnology. More than 50% of the companies consider a global market volume of 250 million EUR or more to be realistic for their product with the highest priority in 2006. Another 26.7% of the companies expect at least a worldwide market volume of 50–250 million EUR. This shows the considerable sales potentials that the companies see.

Considerable sales potential

In this connection, the high hopes placed on Germany's potential market volume in the world market are interesting, too (Fig. 5.38).

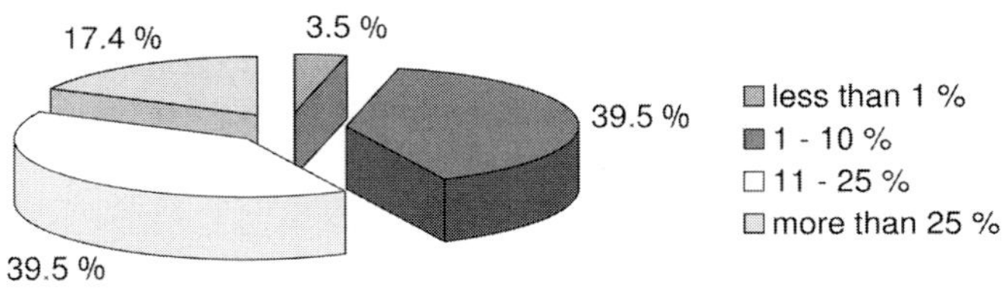

Figure 5.38 Germany's share in the worldwide sales in the selected products of nanotechnology.

Here, it can be seen that Germany's share in the world market in 2006 does not correspond well with Germany's present research potential in nanotechnology. While today the research results are regarded as convincing, the average market shares are between 11 and 25%. Here again it is clear that there is little confidence in the capability of German companies to develop markets.

When analyzing the market shares of their own company in the world market, however, the attitude is more positive (Fig. 5.39).

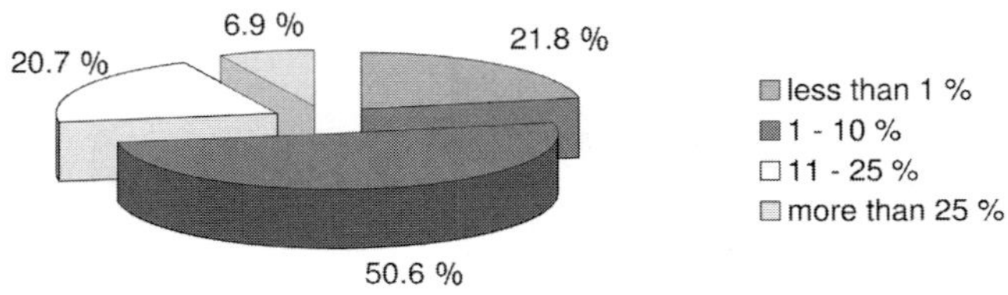

Figure 5.39 Shares of the company in the world market for a specific product.

As far as the product in the application field with the highest priority is concerned, the interviewees see a clear market share for their company.

Evaluation of the interviewees regarding their markets is not only based on their own experience (Fig. 5.40).

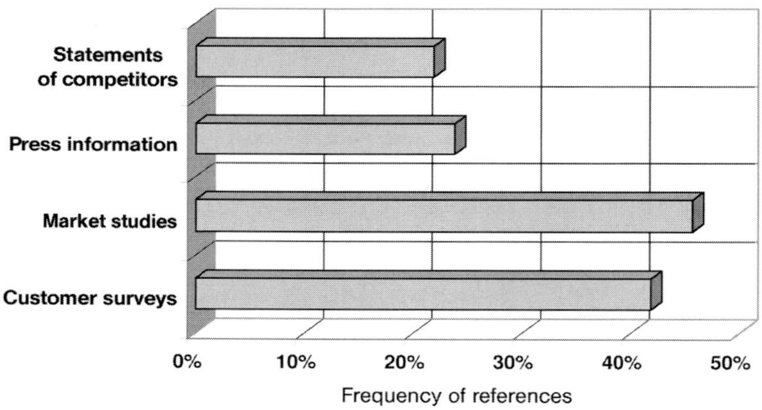

Figure 5.40 Sources of the market evaluation.

A large number of the interviewees preferred to consult third-party information for the evaluation of the market volumes. Press information and market surveys have been analyzed, and to a quite large extent customer surveys have been carried out. This high standing of customer surveys as a source of information shows that the stimulus of user–producer relationships regarding innovations is found in nanotechnology as well.

5.6.2
Obstacles to Successful Nanotechnological Applications in the Year 2006

The interviewees commented in different ways on the obstacles regarding the development and the placement of nanotechnological applications. Here it can be seen that the lack of financial resources necessary for pushing R&D and marketing activities constitutes a central obstacle (Fig. 5.41).

High investment costs as an obstacle

Both investment costs and financing and funding along the value-added chain represent hurdles that exceed any other obstacle. This central problem of financing innovations has a considerable impact on the whole innovation process.

It is interesting to see that the obstacle "legislation" does not play a central role. Obviously, the interviewees do not regard the legislative framework as a primary source of problems. This can certainly be considered as a great success of politics, however, it must be taken into account that the question is aimed at the year 2006, and it may be that the interviewees have not yet anticipated potential frictions within the institutional framework . Here, with regard to nanotechnology a smoother development seems to be possible than in biotechnology, for instance, where the institutional framework already represents a source of continuous conflicts.

Legislation no particular obstacle

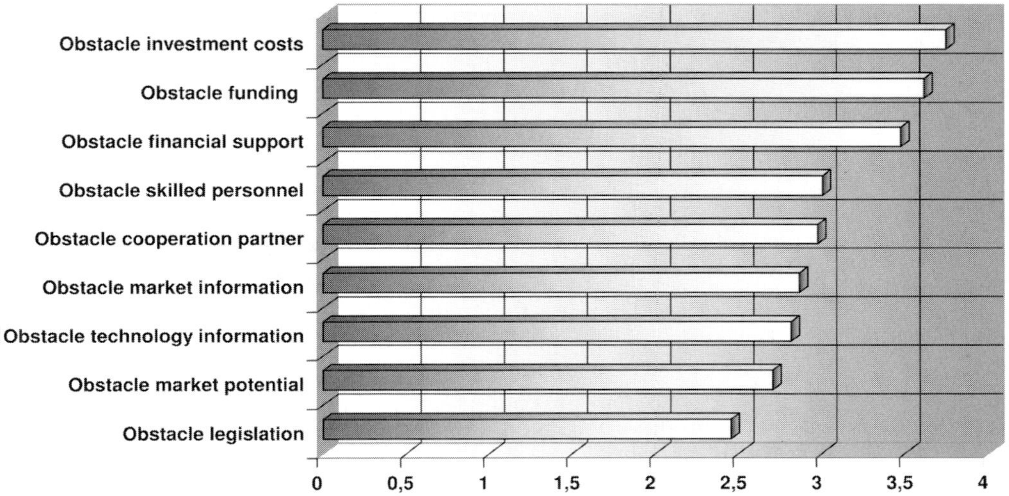

Figure 5.41 Relevance of obstacles to innovation according to the orientation.

5.7
The Importance of Nanotechnology for Different Company Sizes

Characteristics of SME

Within the context of the quantitative survey, an attempt was made to identify statements that differentiate between characteristics of large-scale enterprises and SME. Separate analysis of the quantitative survey by target groups of small, medium-sized and large-scale enterprises led to significantly different results in only a few fields. However, it must be taken into consideration that the sample size for such an analysis is extremely limited. Altogether 34 companies are "large-scale enterprises". In contrast to this, 72 companies can be categorized as "SME". These samples allow only limited statements, and from the statistical point of view they can only be regarded as an indicator, but not as statistically significant. Despite these limitations, seen as an indicator the results are undoubtedly interesting, and extended by the statements of the qualitative survey they are absolutely useful with regard to the innovation policy.

5.7.1
Differences Regarding Business Start-up and Company Funding

Business start-up

It is hardly surprising that there are significant differences between SME and large-scale enterprises regarding time of business start-up and funding. However, looking at this in an isolated way, it is interesting to see that a start-up wave was evident in the year 2000. At least 15.3% of the SME recorded in the sample were founded in this year. Looking at the years 1998 until 2002, the value is almost 35% of the companies recorded in the sample.

Funding

A share of 20.8% of the small and medium-sized companies is financed by means of venture capital. Obviously the New Economy "hype" manifested itself also significantly in venture capital investments in nanotechnology. This is interesting especially against the backdrop of the fact that currently there are hardly any venture capital investments in companies worldwide that are in an early market phase. Therefore, the statement that the access to the capital market is certainly more difficult for young nanotechnology companies today can probably be confirmed.

The share of venture capital companies in young nanotechnology companies lies mainly in the category between 25 and 50%. Here again, it can be seen that venture capital investments are made to such an extent that strategic influence on the company management is possible.

5.7.2
Differences between the Companies' Foci and the Fields of Application Today

Looking at the current foci of the companies' activities, it is noticeable that in the sample it is exclusively SME that already use nanotechnological products and processes for the manufacturing of measuring, control and navigation systems today. This is either attributable to a systematic bias in the random sampling or can be interpreted as an indicator for the growing understanding of companies active in microsystems engineering for nanotechnological opportunities (see Chapter 7 for a detailed discussion).

5.7.3
Differences in Personnel and Employment Trends

Regarding the employment of personnel in the companies, the expected differences between SME and large-scale enterprises are identified in the sample as well.

SME and employment trend

Figure 5.42.makes clear that large-scale enterprises, if committed to nanotechnology, go into this field with a significantly higher number of employees.

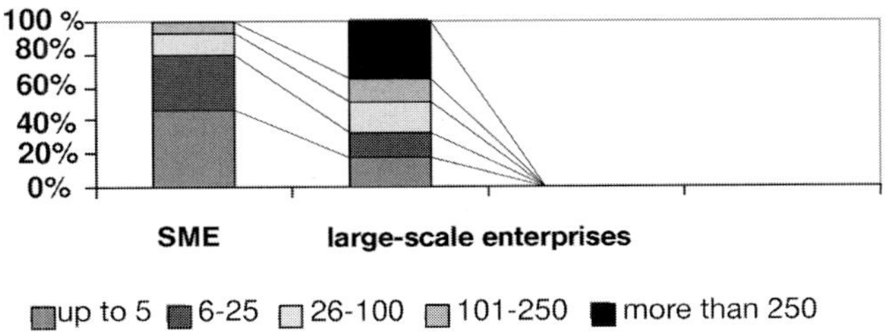

Figure 5.42 Employees in the companies of the sample subdivided into large-scale enterprises and SME.

Nanotechnology has a high potential for the development of new jobs

Both in the case of SME and the large-scale enterprises the predominant number of the companies in the sample intend to increase their number of employees in the field of nanotechnology by 2006. 79.7% of the SME and 84.4% of the large-scale enterprises in the sample expect a relevant increase in employees. With this, nanotechnology has certainly a high potential for the development of new jobs. This is also supported by the statements concerning the expected increase in employees given in the Fig. 5.43.

There are considerable growth potentials for the employment in nanotechnology.

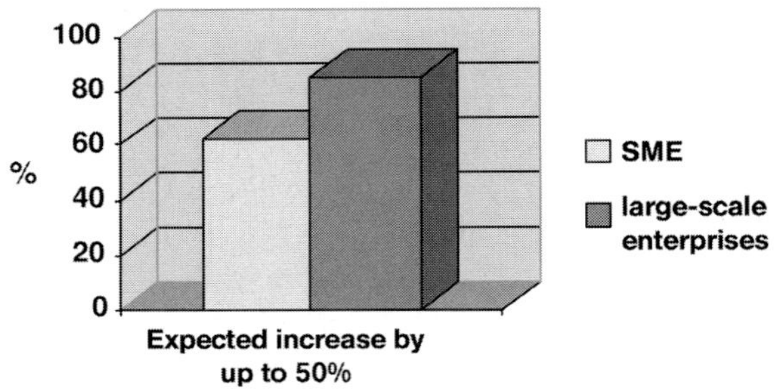

Figure 5.43 Expected increase in employees in the year 2006.

5.7.4
Obstacles to Innovation according to Plant Size

Significant differences between the obstacles to innovation for SME and large-scale enterprises can only be identified in three fields.

SME: Poorer conditions of access to the capital market

There is a clear difference concerning the funding sources. According to their own perception, SME have significantly poorer conditions of access to the capital market than large-scale enterprises. Consequently, 38.6% of the SME see an important innovation obstacle in funding their activities. In contrast to this, the access to the capital market represents an important innovation obstacle for only 7.7% of the large-scale enterprises.

La ck of market information hinders innovation

There is a similar discrepancy regarding access to market information. Here, too, 21.3% of the SME – a significantly higher number than for large-scale enterprises with 3.7% – regard this as an important innovation obstacle.

Finally, the lack of availability of competent regional co-operative partners is another innovation obstacle SME and large-scale enterprises evaluate differently in the sample. 22.2% of the SME regard this innovation obstacle as important, whereas only 7.4% of the large-scale companies consider this to be an important obstacle.

The different plant size classifications reflect the three different opinions of what is seen as an innovation obstacle. Obviously, with increasing plant size, access to the capital market, the available quantity of market information and the attractiveness for regional partners grows as well. For innovation policy, however, which aims at the strengthening of SME, these differences provide a starting point for targeted interventions.

5.8 Overview of the Results of the Company Survey

The central results of the paper-based company survey are summarized below. For a more detailed interpretation of the central results, see Chapter 7.

- The analysis of the business areas of the companies polled shows that companies dealing with nanotechnology are mainly active in the chemical industry and the manufacturing of measuring, control and navigation instruments. This coincides absolutely with the results of other surveys.

Chemical industry very active

- The survey shows that 71% of the companies polled are economically independent and are therefore in a position to set the strategic direction in Germany. At the same time it can be seen that 82% of the remaining 29% are in the hands of German shareholders and only 7% in the hands of non-European shareholders. In about 15% of the companies, venture capital investors are represented.

Independent German companies

- The results of the company survey show that the starting point for nanotechnology lay mainly in the period from 1996 to 2000. Over this period of time, observation of the nanotechnological scene, some R&D work, and exploitation of nanotechnology in products, experienced their sharpest increase. Discussions on market relevance (1996), for instance, and the establishment of competence centers (1998) for the field of nanotechnology with public funding initiated by the BMBF took place within this period as well.

Most frequent starting point

- It is clear that both the number of companies making their first sales with products with nanotechnological shares and the number of companies achieving their total sales with products with nanotechnological shares has increased significantly. Here we should emphasize that 66% of the companies describe themselves as manufacturers and about 29% as purely users of nanotechnological products. Consequently, the group of purely users of nanotechnological primary products has played a less important part so far compared to the active manufacturers of such components. This pattern is typical for a diffusion process. Therefore, it is a good sign for an ongoing diffusion that the number of users acting as purely market observers without achieving sales with nanotechnological primary products decreased from 71% in 1996 to 25% in 2001. Hence, the number of active market players has increased considerably.

Number of active market players increased significantly

- While the analysis of the supply of the primary products suggests that Germany has reached a good position in the field of research, companies in the USA seem to be faster in transforming research results into products (e.g. in the field of I&C). However, it is also interesting to see that currently both Japan and the rest of Asia are

USA and Japan currently better regarding realization

hardly seen as competitors regarding the diffusion of nanotechnology. This result is partly confirmed by analysis of the evaluation concerning the power of potential competitors and R&D and the realization on the market. Here, it can be seen that mainly in R&D, Asia and Europe lag behind the USA and Germany. The interviewees say, however, that regarding the transformation of nanotechnology into products, the USA and Japan are better than Germany.

Positive assessment of nanotechnology by companies

- The majority of the companies assess the prospects of nanotechnology as positive. About 90% of the companies polled intend to enhance their activities in nanotechnology, 30% of them even considerably. This is accompanied by an increase in employment, too. Only 18% of the enquired companies do not see growing manpower requirements for their nanotechnological activities. The rest of the companies (72%) expect strongly growing manpower requirements.

Still concentrated on research

- Currently, German nanotechnology companies are predominantly still concentrated on research, a fact that can be deduced from the priority given to the provision of research personnel, contact to cooperation partners and the real investment activities.

Average shares in the world market

- From the companies' point of view, Germany's share in the world market in 2006 does not correspond to its research potentials in nanotechnology at present. While the research results are regarded as convincing today, the average market shares range between 11 and 25%.

Obstacle investment costs

- Both investment costs and financing and funding along the value-added chain represent obstacles that exceed all other impediments. This central problem of financing innovations has considerable effects on the whole innovation process.

SME characteristics

- Altogether 34 companies of the sample are large-scale enterprises. In contrast to this, 72 companies can be categorized as SME. 35% of these SME were established between 1998 and 2002. A significant difference can be seen in the funding sources. In their perception, SME have significantly poorer access conditions to the capital market than large-scale enterprises and their access to market information is more complicated.

6
Market Potential of Nanotechnology in Lead Markets

Wolfgang Luther, Andreas Hoffknecht, Dirk Holtmannspötter, Norbert Malanowski and Volker Wagner

6.1
Notes on the Procedure

All the edited results of the literature analysis, patent analysis, expert interviews and the company survey regarding the economic potential of nanotechnology were brought into expert workshops in the form of propositions. These formed a basis for constructive confrontation using the Delphi method (described in Chapter 1 Footnote 2). Experts from the respective Lead Market to be surveyed (see Section 6.2) were questioned about the topics:

Method: expert workshops

- nanotechnological product/market potentials along the value-added chain;
- innovation/implementation obstacles;
- employment trends/qualification and
- patents.[1)]

A list of questions which was used to guide the expert interviews is shown in Appendix 1. The participants in the workshops were presented with consolidated results and propositions that they should analyze and review critically with regard to their realistic evaluation. Moreover, nanotechnological product options were reviewed and partly supplemented using a Metaplan-based card questionnaire.

In the workshops, 15 to 25 experts from banks, science, competence centers for nanotechnology, producers, suppliers, system developers and venture capital companies took part, dealing with the subjects of chemistry, car manufacture and optics.[2)] A central problem that might occur in workshops using the Delphi method is the risk that the chosen experts may try to present their special subject areas in a particular way (Heimer and Werner, 2004). This possible problem was avoid-

Experts from relevant fields of operation

1) For reasons of clarity, the results on the subject of "patents" are to be found in Chapter 4.

2) Bilateral discussions on the results of this study were held with the experts from the European Patent Office.

Growth Market Nanotechnology. Edited by N. Malanowski, T. Heimer, W. Luther, and M. Werner

ISBN: 3-527-31457-1

ed by using a large number of experts from the relevant fields of operation and ensuring that an adequate balance was maintained between the fields of operation.

In order to demonstrate the importance of nanotechnology for the German Lead Markets and to show the prospects of the development of market opportunities, the results are correlated with the respective lines of industry and depicted as such. These results are found in the respective sections of Chapter 6.

6.2
Opening-up of Sales Opportunities in Lead Markets

Lead Markets are of particular economic and strategic importance

Usually, in more general surveys on the economic potential of nanotechnology the respective so-called Lead Markets[3] of the individual countries are not sufficiently taken into account – as mentioned in Chapter 1.1. While, for example, the fields of electronics, I&C and biotechnology are Lead Markets in the USA, in Germany the Lead Markets are, inter alia, the chemical industry/process engineering and car manufacture. A characteristic of these Lead-market industries is their particularly intensive partnership with science, from which they draw their technological power, and the fact that they are usually of special economic or strategic importance for a country.

In the early 1990s Michael Porter, a leading Harvard economist, described a demand structure that stimulates the faster development of an innovation on a national or regional market and sets trends that cause the acceptance of the innovation on other markets as "the dynamic of the so-called Lead Markets". Companies can carry out market research and R&D in a Lead Market to test and develop their products in a demanding environment. Governments and other national institutions can promote their country and safeguard and develop high-quality jobs and corporate divisions (Krück et al., 2002).

To demonstrate the importance of nanotechnology for German Lead Markets and the prospects for the opening up of sales opportunities, special attention was directed to such fields of industry that make up the technological efficiency of German industry (chemistry, car manufacture, optics, life sciences[4] and electronics[5]).[6] Before their final documentation, the data gathered by such a concentration

3) See Chapter 1.1 for the definition of Lead Markets.

4) In a study by AGIT published in autumn 2004, results regarding the market potentials of nanotechnology in the field of medical technology/life sciences are to be found. To avoid duplication, a workshop was not organized in this subject.

5) As already mentioned in Chapter 1.2, electronics is a technology whose future development is described as very good, thanks to the International Technology Roadmap for Semiconductors. Since the semiconductor industry is very investment-intensive, market forecasts are extremely important for entrepreneurial decisions.

was edited by the project team again and fed into the industry-specific network for validation.

6.3 Chemistry

In Germany, the chemical industry is traditionally one of the businesses with the strongest export activities; more than half of the turnover achieved by Germany is achieved directly in foreign countries. Global market growth is significantly higher than the domestic growth. Extremely high import and export quotas compared to all processed industrial goods are an indication of the strong international integration of key chemistry and special chemistry into the mutual exchange of goods. Key chemistry goods reach about 70%, while special chemicals show values of 90% (VCI, 2003).

Strong internationalization in the chemical industry

Within key chemistry, Germany has advantages due to specialization in the fields of dyestuffs, inorganic matter and synthetic materials. With an innovation activity clearly above the average of the manufacturing sector, the chemical industry has considerable influence on other industries, providing them with primary products and ideas for innovations. There is no better proof for its key function than this close economic interconnection. Key chemistry and special chemistry are branches that meet with particularly favorable demand conditions in Germany: up-market customers (car manufacture and mechanical engineering, medical technology and analytics) force up a high investment rate. Important customers, for instance, were the car industry, the packaging industry and the building industry. Chemical products are also applied in the fields of health, environment and nutrition, and contribute to a large extent to an improvement in quality of life. The largest share in the production value of Germany's chemical industry was held by the fine and special chemicals with a good 25% in the year 2002. Polymers and pharmaceutical products followed.

The German chemical industry has an unchallenged top position regarding R&D in the international comparison. Among all the other Western industrial countries, only Japan is ranked before Germany. Even the chemical industry in the USA produces in a less R&D-in-

This requirement is fulfilled by a lot of (commercial) market research institutes. Currently, the forecasts reach approx. to 2008. How the ratio of nanoelectronics to microelectronics is going to develop and which consequences will arise for which industry, has not been surveyed systematically enough yet, however, the results could be filtered out by means of secondary analysis of a selection of commercial market surveys. For this reason, the organization of a separate workshop was not required.

6) See, inter alia, the reports of the BMBF "On the technological efficiency of Germany" since the end of the 1990s.

tensive way than the chemical industry in Germany does. There is a trend indicating that chemical research seems to become more and more the business of large-scale enterprises (BMBF, 2002).

Within the German economy, the chemical industry is ranked particularly high: the share of chemical companies in the turnover of the manufacturing sector is about 10%, ranking fourth among the industries of the manufacturing sector. With an average annual growth of 1.8% between 1991 and 2002, the chemical production was growing faster than the production of the manufacturing sector altogether. The expenditure for the maintenance and development of the technological basis are immense: with 8.2 billion EUR, 18.2% of the R&D expenditure of the German economy is borne by the chemical industry. In 2002, Germany's chemical companies employed on average 462 000 people in Germany. With this, this line of business ranks sixth among the industries of the manufacturing sector. The positive employment effect of the chemical industry goes far beyond its own line of business. Owing to the demand of the chemical companies, domestic suppliers create more than 380 000 jobs.

The chemical industry produces a wide range of products for many different areas of life. It manufactures primary products for other industrial branches. Inorganic basic chemicals, petrochemicals, polymers as well as fine and special chemicals belong to this group. In the year 2002, altogether about 70% of the chemical production was supplied to industrial processing companies

Generation of high-quality special chemicals by nanomaterials

In future, nanotechnology, especially the field of nanomaterials, will play a more important role for the chemical industry in the generation of high-quality special chemicals mostly at the value-added stage of primary or intermediate products. In some sections, the application of nanomaterials has been established for a long time, for instance, in the case of industrial carbon black, pigments, polymer dispersions and colloids. In other sectors, especially for application in the health system or electronics, a lot of new nanomaterials are in the stage of research and will only develop their economic potential in the years to come. Since material innovations give an essential impetus to technological progress, new nanomaterials will be the pacemakers for innovative products in other branches of industry too.

6.3.1
Value-added Chains and Application Potentials

Nanomaterials, precursors and coating materials are the base of the value-added chains of nanotechnology in the field of chemistry. From these elements, primary products and semi-finished products are manufactured in the next step, which are processed both in the chemical industry and in other branches of industry. The final applications

Table 6.1 Potential applications of nanomaterials in different stages of the value-added chain

Elements (nanomaterials/precursors)	Primary products semi-finished products	Applications/Products (components/systems)
Inorganic nanoparticles metal oxides, layer silicates, fullerenes, CNT, carbon black, aerosil, lanthanoid **Organic nanoparticles** polymer dispersions, coloring substances, active substances, macromolecules (dendrimers etc.) **Nanoporous materials** aerogels, zeolites **Nanocomposites** glass/ceramics, metals/alloys, polymers, organic semiconductors, ferrofluids etc. **Precursors/coating materials** metalloorganics, various chemicals, PVD targets	• atalysts • membranes/filters • pigments/dyes • polish • fillers • sensors • active substances/carriers • films/packaging • textile fibers • marker materials • supraconductors • thermo-electrics • coated semi-finished products	**Medicine** drug delivery, biochips, implants, antimicrobial material **Cosmetics** suncream, toothpaste **Automobile** tires, bodywork materials, exhaust emission purification fuel cell, windows and mirrors, LED lighting **ICT** data storage systems, displays, laser diodes, glass fibers **Energy/Environment** solar cells, batteries, fuel cells, condensers

and chemical products include nearly all branches of industry, of which especially medicine, cosmetics, car manufacture, I&C technology as well as energy and environmental technology are the economically most important for nanotechnology (see Table 6.1).

Within the framework of a workshop, representatives of German nanotechnology companies in the chemical industry prioritized the economic potential of different nanotechnological applications for the time horizon 2006. The results are given in Figs. 6.1, 6.2 and 6.3.

Intermediate products of the chemical industry with integrated nanotechnological applications can be subdivided into the following categories:

- Catalytic converters (nanoporous ceramic catalyst carriers, nanoparticular stainless steel catalytic converters for a variety of chemical reactions and processes etc.)
- Filtration and separation (nanoporous membranes based on organic, inorganic or organic-inorganic hybrid materials)
- Composite materials (nanoparticle-reinforced polymers, ceramics or metals)
- Coating materials, (brine and dispersions for the coating of metals and synthetic materials)

- Additives/fillers (nanoparticular fillers such as silicic acid, carbon black as additive for a variety of chemical products such as rubber, pharmaceutical products, dyes etc.)
- Micronized active substances (nanoparticular organic active substances e.g. vitamins as food additives or pharmaceutical active substances in drugs, inter alia, for a better bioavailability)
- Drug carriers (nanoparticular transport systems for pharmaceutical active substances for the selective transport of active substances or nanoporous materials for a controlled delivery of active substances, e.g. for fertilizers or pesticides in farming, aromatic substances in textiles etc.)
- Dyes/lacquers (scratch-resistant clear lacquers based on polymerization, ormocers, dendrimers etc, effect colors based on monodisperse nanoparticles etc.)
- Sensors (temperature, pressure and chemical sensors based on nanomaterials, e.g. metal oxides, nanotubes or nanostructured graphite)
- Adhesives (polymer dispersions, adhesives based on magnetic, nanoparticular composites for switchable adhesives, metal nanopowder as additive for conductive adhesives etc.). Moreover, there is a multitude of possible applications in other branches of industry (see Chapter 3.2).

In a workshop with representatives of German nanotechnology companies active in the field of chemistry/coating technology, nanotechnological products were evaluated regarding their economic relevance until the year 2006. The results are broken down according to the fields of key chemicals, intermediate products and final applications and are depicted in Figs. 6.1, 6.2 and 6.3.

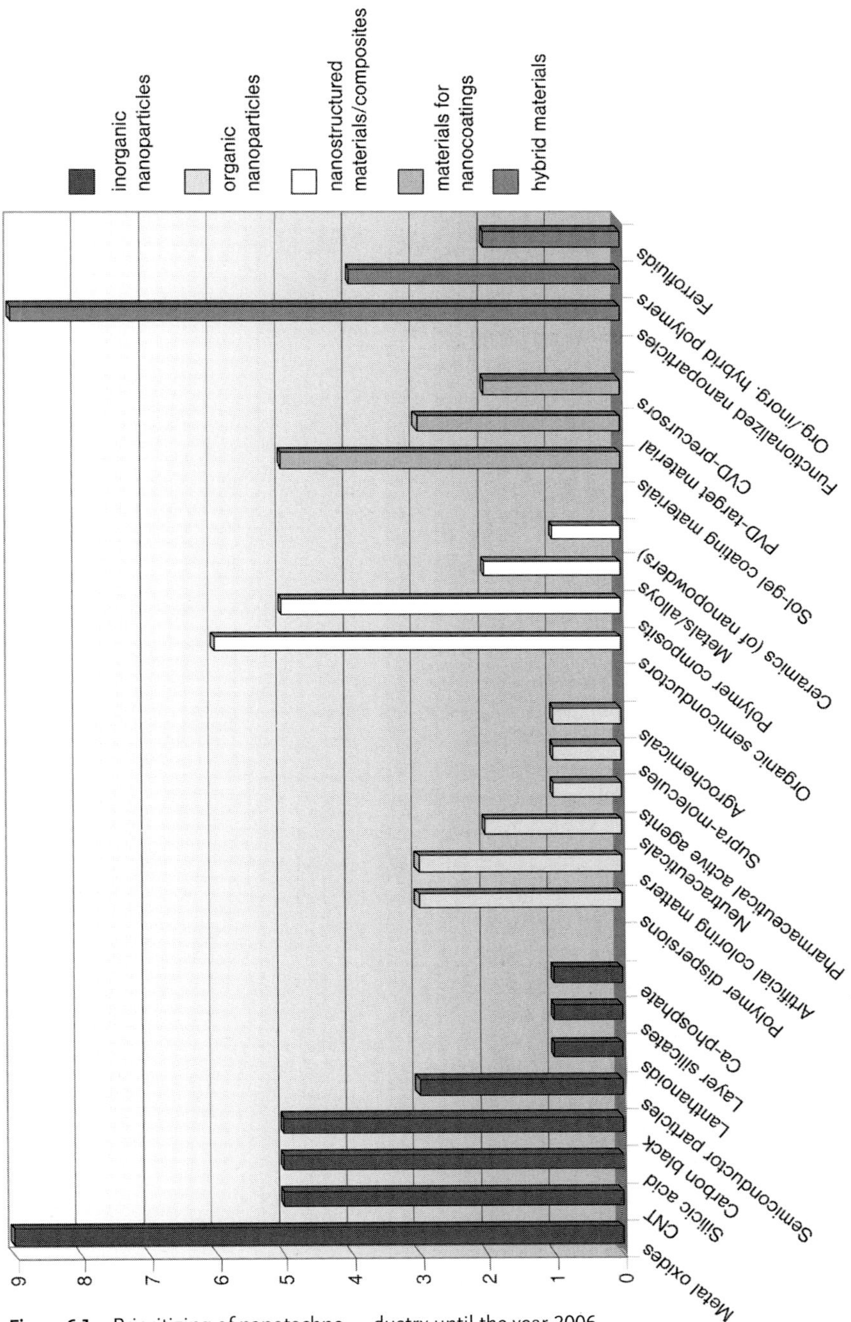

Figure 6.1 Prioritizing of nanotechnological key chemicals regarding the market relevance for the chemical industry until the year 2006 (the number of references given by the participants of the workshop is stated).

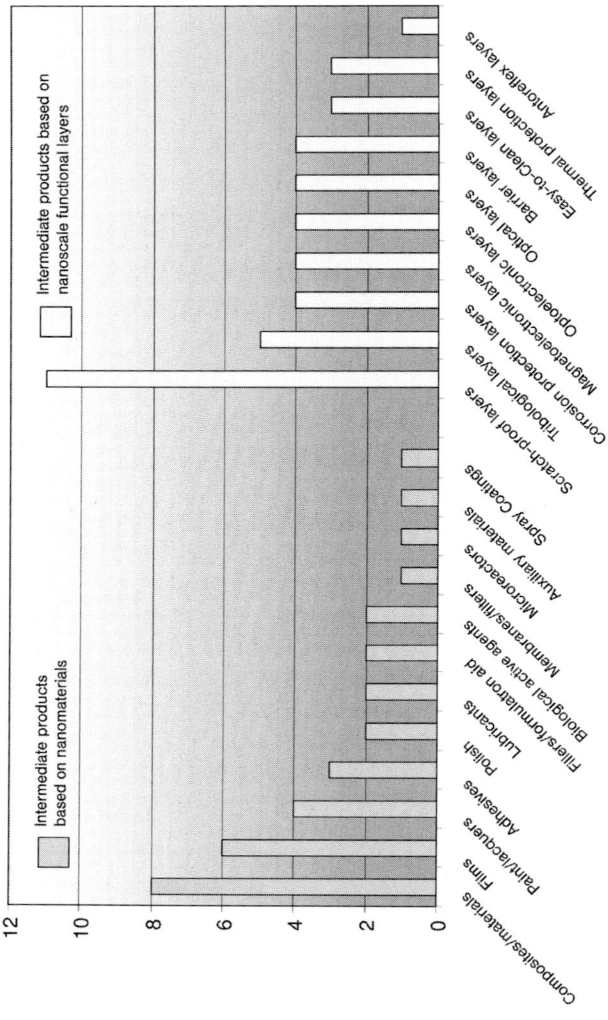

Figure 6.2 Prioritizing of nanotechnological intermediate products regarding the market relevance for the chemical industry until the year 2006 (the number of references given by the participants of the workshop is stated).

The highest relevance for the economic potential of the chemical industry until the year 2006 was attributed to:

Key chemicals:
- Metal oxide nanoparticles (e.g. titanium dioxide)
- Functionalized nanoparticles
- Organic semiconductors

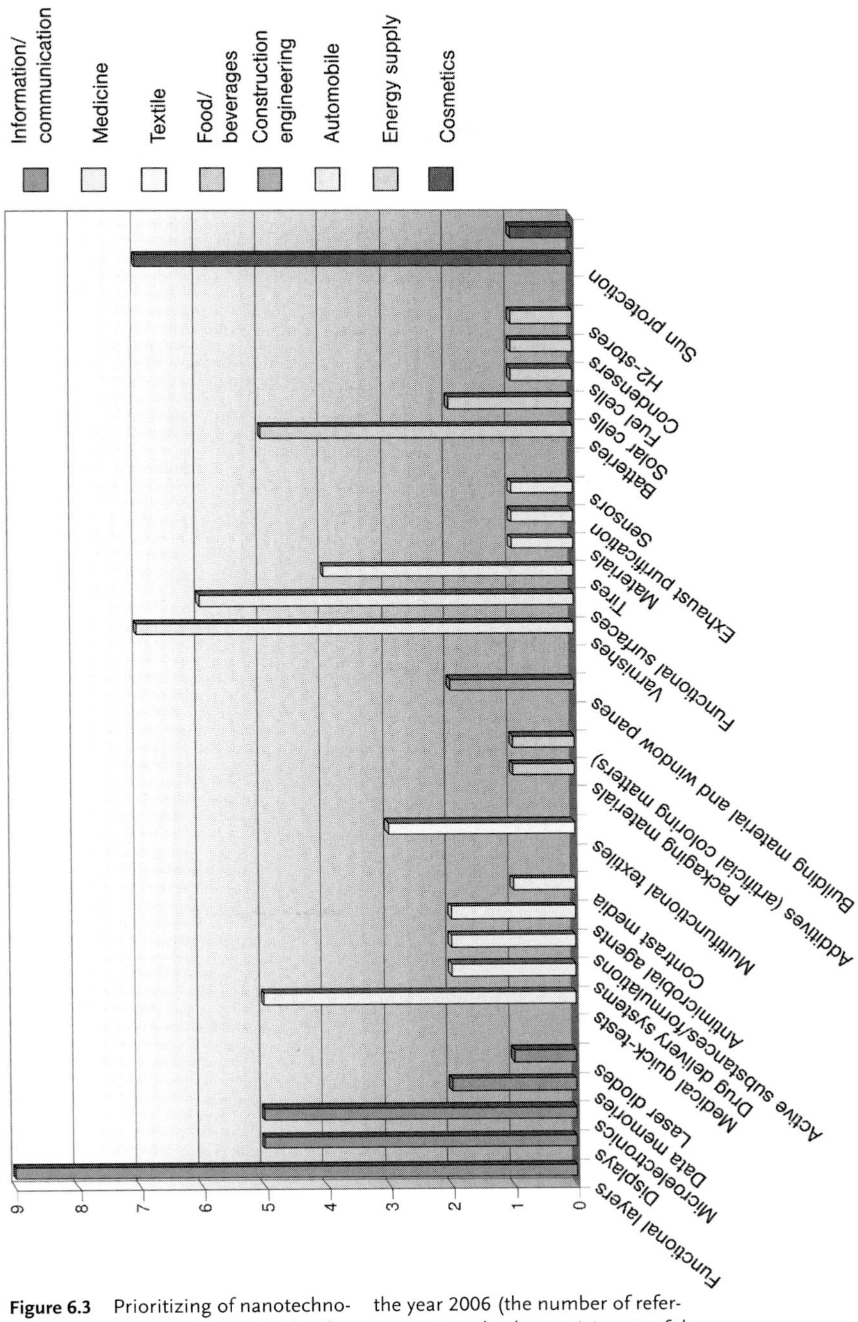

Figure 6.3 Prioritizing of nanotechnological products in different fields of application regarding the market relevance for the chemical industry until the year 2006 (the number of references given by the participants of the workshop is stated).

Intermediate products of the chemical industry:

- Nanocomposites
- Scratch-resistant coatings

Products in other fields of application:

- Information and communication (electronics based on functional layers)
- Automobile (lacquer and functionalized surfaces)
- Cosmetics (sun-protection products based on metal oxide nanoparticles)

In the following, value-added chains in the chemical nanotechnology for the examples regarded as economically relevant within the framework of the workshop are explained in more detail. Table 6.2 summarizes information on applied nanomaterials, on intermediate products, applications, technologies and the product features improved by nanotechnology.

Table 6.2 Description of the value-added chains for different prioritized nanotechnological applications in the chemical industry.

Nanomaterials	Intermediate products	Applications	Technologies	Improved product features
Metal oxides	Dispersions, sols, CMP-slurries	Electronics, sun protection, diagnostics, transparent conductive layers	Manufacturing from vapor/ liquid phase, precipitation	Transparency, workability, homogeneity
Functionalized nanoparticles (coated nanoparticles, overcured polymers)	Composites, dispersions layers, varnishes, films, biomarkers	Product/ plagiarism protection (documents, luxury goods), printable electronics displays, diagnostics, automotive paints, packaging	Manufacturing from vapor/liquid phase, precipitation, surface modification, processing through ink-jet	Mechanical effect (composites), scratch resistance, self-cleaning, high spec. surface (additives), transparency, workability
Organic semiconductors	Optical layers, electronic layers	Lighting, displays, flexible electronics, sensors	Manufacturing from vapor/liquid phase	Optoelectronic and semiconductor features, workability
Carbon nanotubes	Polymer composites, membranes, fibers, field emission tips	Electronics, displays, design material, antistatic packaging	Deposition from the vapor phase (CVD)	Conductivity, mechanical properties, field emission properties

6.3.2 Market Potentials

Market potentials in nanotechnology for the chemical industry result mainly from the manufacturing of nanotechnology-based key and intermediate products that are partly processed in the chemical industry, but also in other branches of industry. In Table 6.3 the assessments of the market potentials for different nanostructured key and intermediate products are shown. The depiction of the market potentials of the end products was excluded from this table, as the value added arises mostly outside the chemical industry and even beyond the nanotechnology sector.

Table 6.3 shows that especially the "classical" nanomaterials, such as carbon black, polymer dispersions or zeolites currently show annual turnovers in the range of billions of US dollars. However, for more recent nanomaterials, such as nanoparticles, carbon nanotubes and polymer nanocomposites, an annual market growth of several 100% is expected in the mid-term. These assumptions, however, are based on very optimistic conditions, for example supposing that manufacturing processes for nanomaterials are scalable to an industrial standard without problems, and that considerable price-performance profits are possible through nanomaterials.

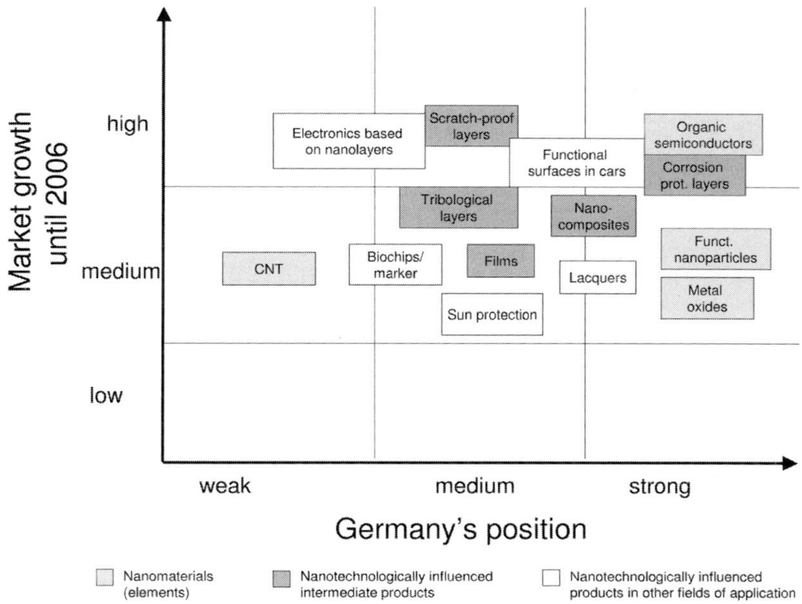

Figure 6.4 Market growth and Germany's position in the field of nanotechnological products of the chemical industry.

Table 6.3 Assessments of the world market for nanotechnological products in the chemical industry.

Key/intermediate products	Annual world market volume (reference year)
Raw materials	
CVD precursors	50–250 million EUR (2006)[12]
Sol–gel materials	50–250 million EUR (2006)[12]
PVD targets for magnetoelectronics	300 million USD (2006)[13]
Nanomaterials	
Metal oxide/metal nanoparticles	900 million USD (2005)[1]
Nano silicic acid	800 million EUR (2003)[2]
Nano-layersilicate	25 million USD (2006)[3]
Carbon nanotubes	145 million USD (2005)[4]
	1.2 billion USD (2006)[5]
Carbon black	3 billion USD (2002)[6]
	5.7 billion (2002)[3]
	8 billion USD (2006)[5]
Polymer dispersions	15 billion EUR (2002)[7]
Organic semiconductors	500 million USD (2005)[10]
Dendrimers	5–15 million EUR (2006)[3]
Micronized active substances (vitamins, pharmaceutical products)	1 billion EUR (2002)[7]
Zeolites	2.6 billion USD (2006)[5]
Aerogels	10 billion USD (2005)[8]
Polymer nanocomposites	300 million USD (2006)[3]
	1.1 billion USD (2006)[5]
	1.5 billion EUR (2009)[9]
Intermediate products	
Corrosion protection paper	10–50 million EUR (2006)[12]
Lacquers	50–250 million EUR (2006)[12]
Films for displays	50–250 million EUR (2006)[12]
Marker materials	250–500 million EUR (2006)[12]
Nanosensors	
Temperature sensors	4.6 million USD (2004)
	217 million USD (2011)[11]
Pressure sensors	4.4 million USD (2004)
	87 million USD (2011)[11]
Chemical sensors	1.3 million USD (2007)
	36 million USD (2011)[11]

1 BCC, 2002, **2** Wacker Silicones, 2003, **3** SRI, 2002, **4** Mitsubishi Research Institute, 2002**5,** Fecht et al., 2003, **6** Reuters, 2002, **7** BASF/Distler, 2002, **8** Aspen Systems, 2001, **9** Stevenson, 2003, **10** Frost&Sullivan, 2002, **11** Frost&Sullivan, 2003, **12** company survey **13** Platinum Association, 2003.

In the opinion of German nanotechnology enterprises of the chemical industry, growth potentials are mainly seen in functional coatings (e.g. electronics, scratch-resistant layers, corrosion protection, surface functionalization of automobile components), but also in the field of organic semiconductors and nanocomposites. German companies have reached a strong position in the world market, particularly in the fields of functionalized nanoparticles, metal oxide nanoparticles and organic semiconductors. In the field of carbon nanotubes, where only a few German companies are represented, Japanese (e.g. Mitsui) and American companies (e.g. Carbon Nanotechnologies) are dominant.

6.4 Car Manufacture

Germany has a leading position in car manufacture worldwide, owing mainly to the fact that the focus was concentrated on high technology in order to be able to fulfill the increasing demands of the markets and to maintain the export power of German car manufacturers. The car industry, which represents a macroeconomic key industry for Germany, made a turnover of 202 billion EUR in 2001. 69% was attributable to manufacturers of motor vehicles and motors, a good 3% to the manufacturers of trailers, body work and containers; 28% to the car components and accessories industry. With this, the car industry made one sixth of the total industrial turnover in Germany. With 121 billion EUR, the German car industry earned 60% of its sales in foreign countries. German car manufacturers and suppliers employ more than 1.5 million people worldwide, more than half of them in Germany. Recently, domestic employment has been expanded: Since 1994, the number of jobs at home has increased by 122 000. Despite economic weakness in 2002, employment in the car industry remained almost constant with 763 500 people. Owing to the assumption of additional value-added activities from the manufacturers, suppliers even employed additional staff. The German car industry employs more than 12.8% of the working population of the whole German industry. At the same time, it raises one third of the R&D expenditure of the German economy and one fifth of the investments. During the last five years, 49 billion EUR were invested in Germany; over this period of time, expenditure for R&D exceeded 65 billion Euro, and the R&D staff was increased significantly to more than 70 000 employees (VDA, 2003).

High innovation pressure will encourage the application of nanotechnology in car manufacture

Consequently, innovation and high technology are compelling conditions for the maintenance of competitiveness in car manufacture. Increasing government regulations regarding safety and environmental compatibility as well as steadily growing expectations of the customers concerning power, comfort and design of cars will be a constant impetus for the introduction of innovative technologies.

These basic conditions will certainly boost the diffusion of nanotechnological applications in car manufacture. Owing to its broad cross-sectional character, a large number of automotive technologies will be influenced by nanotechnology. It is clear that nanotechnological competence in future car manufacture will be one of the core abilities that are required to maintain the international competitiveness for this industry so important for the German economy (TAB, 2003).

6.4.1 Value-added Chain and Application Potentials

Nanotechnology will enable progress with regard to all criteria relevant to the further development of the automobile, from ecology (with the aspects of energy-efficient drives, light-weight manufacture, emission control and protection of resources) to safety (passive and active safety) and comfort (with the aspects of passenger wellbeing, product design and infotainment). However, this will only be possible against the backdrop of economically competitive technologies with a competitive price-performance ratio (see Fig. 6.5).

Today, the maturity stage of nanotechnological developments in car manufacture ranges from currently used components or partial systems (e.g. non-reflective instrument coatings) to concrete development efforts (e.g. anti-fog windows) up to the stage of predicted product ideas with, at best, long-term feasibility (e.g. switchable gloss paints or self-designing bodywork). Nanotechnological applications

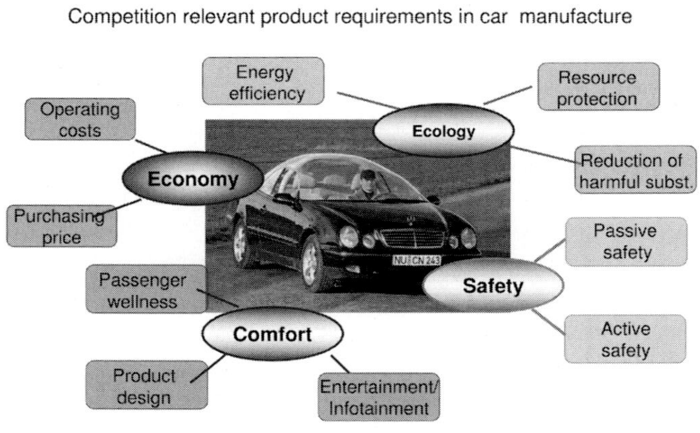

Figure 6.5 Competition-relevant product requirements in car manufacture (Source: Germany's car manufacturers, VDI TZ).

Table 6.4 Potential nanotechnology applications in car manufacture in different stages of the value-added chain.

Key elements/processes (nanomaterials/tools)	Components/systems	Applications
Materials • High-tenacity steels • Metal matrix composites • Nanoparticle-reinforced polymers • Catalytic nanoparticles • Thermoelectrics • Nanoadhesives • Nanofluids **Processes/Tools** • PVD-, CVD-process • Nanoparticle synthesis • Ion beam/ Plasmaprocess **Functional layers** • Ultra-hydrophobic • Electrocromic • Anti-reflective • Wear protection • Scratch resistance **Electronics and sensor technology** • Magnetoelectronics • WBG-semiconductors (SiC, GaN) • LED, OLED	**Chassis** • Tires • Shock absorber • Sensor technology **Drive system** • Ignition system, injection • Fuel tank additives • Exhaust system • Fuel cells • Batteries • Lubricants, cooling • Thermo-electric waste heat utilization **Bodywork/outer skin** • Support structure • Windows • Paint work • Outer skin functionalization **Interior/fittings** • Consoles/fittings • Displays • Lighting technology • Electronics, DP	**Safety** • Active safety (brakes) • Headlights, visibility (indirect visibility, radar etc.), road behavior) • Passive safety (car structure, airbag, pedestrian safety **Comfort/design** • Air-conditioning • Effect paintwork • Self-cleaning surfaces • Entertainment/ infotainment (internet, video services) • Navigation systems/traffic guidance systems **Environment/lastingness** • Fuel consumption • Car exhaust emission • Noise emission • Resource saving production • Recycling

are discussed for a large number of product groups and components, as follows in Table 6.4.

Some essential applications of nanotechnology in car manufacture are briefly explained below (see FHG-INT, 2003 or rather TAB, 2003).

6.4.1.1 Nanomaterials as Structural Materials

In the field of structural materials, nanotechnological development efforts are mainly aimed at the reduction of the vehicle weight using light-weight materials to enable fuel savings. This concerns, for example, nanoparticle-reinforced polymers or polymers with nanofillers, the physical and chemical features of which can be modified and optimized by these fillers. All in all, a great future is predicted for these polymer-based nanocomposites, particularly in car manufacture, provided that, apart from solving the outstanding technological problems, it is possible to reduce the prices that are still very high.

Application possibilities occur wherever conventional synthetic materials have been used up to now, hence in the whole sector of interior fittings, but for example, even in casings for electric components (Hohenberger, 2000). The properties of metals can be improved by the integration of further phases in the nanometer range; so the Japanese steel manufacturer NKK is developing nanoparticle-reinforced steels for the light-weight construction of bodywork (Materials World, 2001). The mechanical properties of other metals, such as aluminum relevant for car manufacture, can be enhanced specifically by means of nanostructured ceramic fibers (especially by silicon carbide, but also by aluminum oxide or aluminum nitride). Another important matter is the issue of corrosion protection for the maintenance of the functions of the materials used for design purposes. Here, nanotechnology offers solutions, for instance through the application of sol–gel-based corrosion protection layers with inorganic–organic hybrid materials which represent an environmentally friendlier alternative to the heavy-metal-containing corrosion protection systems and which are also more competitive from an economic point of view. In Germany, for example, Nano Tech Coatings (NTC) is dealing with the manufacturing of such corrosion protection systems (Aschenbrenner, 2003).

Nanomaterials for light-weight construction, corrosion and wear protection in car manufacture

Wear protection of mechanically highly stressed components, such as diesel injectors for example, is another field of application of nanotechnology in car manufacture. Here, nanostructured hard coatings on carbon (DLC), carbide or nitride basis produced by plasma techniques are the adequate means to achieve higher injection pressures necessary to meet the growing requirements regarding fuel savings and emission control. In Germany, such systems are being manufactured, inter alia, by Bosch (Bosch, 2001). The current DLC-applications in the motor vehicle sector, such as the coating of highly stressed components of diesel injection systems or friction and wear-reduced valves, use layer thicknesses in a range of a few µm with a tendency to further reduction.

Furthermore, weight reductions are possible through the application of synthetic material instead of glass in car windows. Transparent nanoscale layers can give such plastic windows, for example on polycarbonate base, the scratch resistance of mineral glass, and thus enables their utilization in highly stressed car windows. In Germany, these developments are pushed ahead inter alia by the Exatec GmbH, a joint venture of Bayer AG and General Electric. The appropriate coatings have already been used for plastic lenses for some time. In the long run, the aim is the development of transparent light-weight materials with low density and high stiffness which could enable a transparent domed roof and would lead to an improved all-around visibility when A, B and C pillars are not needed.

Particularly important in the engineering sector are also nanotechnological gluing techniques and adhesion promoters which provide energy savings in joining processes (gluing instead of welding), the replacement of environmentally harmful adhesives and the simplification of recycling processes. An interesting application could result from adhesives modified with magnetic nanoparticles. The bonding properties of these adhesives are controllable from outside by electromagnetic radiation with the integrated thermal energy generating chemical reactions or a thermal bonding. Such applications are developed by the Sustech GmbH in Darmstadt.

The utilization of nanostructured carbon black particles as fillers in car tires has been state-of-the-art for a longer time. Progress in the manufacturing of new types of nanostructured carbon black, inter alia from Degussa, will enable the further optimization of tire qualities, such as the reduction of the rolling resistance.

6.4.1.2 Surface Functionalization

An important field of application of nanotechnology in car manufacture is the functionalization in the field of car surfaces directly perceptible by the user, as for example the paintwork, windows or instruments in the cockpit. Here, a lot of ideas and partly realized applications exist (see FhG-INT; 2003, TAB, 2003).

Broad field of applications for nanostructured functional surfaces in car manufacture

A standard application already practised for a long time is the manufacturing of headlight reflector coatings. Both the wafer-thin barrier primary coating of the reflector and the vapor-deposited aluminum reflection layer as well as the corrosion protection surface is only nanometer thin. The essential advantages in contrast to conventional coatings are the higher reflection ratio, and with this a higher light brilliance as well as longer durability owing to the especially high density of the corrosion protection layer (Rügheimer and Schiller, 2002). Another application already in use is anti-reflection coatings on cover plates in the display sector. This application of nanoscale interference coatings long established in industry could be further developed in future, particularly with regard to more economical production methods and to coating large, even curved surfaces. Here, inter alia, approaches to the solution are the optical interference coated sol–gel-layers applied by means of dip-coating and offered, among others, by Schott.

Other fields with a high product potential in car manufacture are anti-fog layers and easy-to-clean-layers by hydrophilic respectively hydrophobic surfaces.

Anti-fog and self-cleaning surfaces

Anti-fog layers are viable through the application of a chemically modified hydrophilic nanolayer. Possible applications are seen in the coating of the insides of windows, outside mirrors or headlight cover plates. There are a lot of development activities and prototypical applications. Mass-production, however, is still not on the horizon due to the high costs for an insufficient abrasion resistance and the hardness of the layers (Langenfeld et al., 2001). Hydrophobic surfaces that can be produced with the help of various process technologies by means of applying non-polar nanolayers (especially through carbon-fluor-compounds) are discussed in connection with easy-to-clean or self-cleaning surfaces (e.g. the "lotus effect") in a wide range of application fields. In car manufacture, the respective equipment of outside mirrors, paintwork surfaces or the outside of the windshield are regarded as possible applications. Comparable to the hydrophilic systems, current obstacles to a further diffusion in car manufacture are high costs, insufficient mechanical properties as well as low long-term stability. Nevertheless, first products are already on the market, for example from Nanogate.

Other market-relevant applications of nanotechnology in car manufacture are effect varnishes based on nanoscale pigments. These are based on the application of nanoscale plates of silicon or aluminum oxide coated with nanoscale interference layers of metal oxides, for instance of titan or ferric oxide, that enable the feasibility of different optical effects (Sepeur-Zeitz, 2003).

6.4.1.3 More Energy-efficient Drives and Energy Supply

A large number of development efforts and concept proposals for the utilization of nanotechnologies in cars concern the field of energy supply/drive. Among others, the following fields are to be mentioned:

Solar cells

Solar cell modules integrated into the outer skin of the car, for example, could allow the operation of air-conditioning systems even when the engine is switched off, which would entail considerable fuel savings. Various nanotechnological approaches pursue the optimization of photovoltaic systems suitable for utilization in car manufacture, such as Thin-layer solar cells on the basis of polymers (e.g. Grätzel-type) or silicon (e.g. Si-Ge-quantum dots).

Batteries/condensers

The utilization of nanomaterials as electrodes enables the development of more powerful batteries and condensers for future applications, e.g. in cars with hybrid drives.

Fuel cells

Nanotechnologies play an important role in the optimization of fuel cells, mainly in the extension of reactive surfaces of electrodes, membranes and catalytic converters. Nanomaterials are also considered for efficient hydrogen storage, for example nanocrystalline light metal compounds or even metal organic compounds (see Müller, 2002).

Thermoelectrics

The application of thermoelectric converters for the utilization of thermal energy on hot modules like cylinders in the motor of a car could be of increasing importance in future. The efficiency of thermoelectric converters can be improved by using nanocrystalline materials.

6.4.1.4 **Electronics und Sensor Technology**

In the field of car electronics and sensor technology, nanotechnology arrived a long time ago and still has a considerable future potential, among others, in the following fields of application (see Dressen, 2003):

LED

LED tail lights and stop lights are already state-of-the-art. They owe their working life and light efficiency at minimal power consumption to a layer system of semiconductor materials with nanometer precision. Further optimizations could even make LED-headlights possible within this decade. Because of the LED lights being much smaller than conventional car lights, car manufacturers benefit from a larger scope of design, for example for the further reduction of air resistance and thus of the fuel consumption. In the end, the considerably lower energy consumption of the LED is also for the benefit of the reduction of the fuel consumption.

Sensors

The heavily increasing number of sensors in car manufacture, as required for instance for motor control, room climate and for safety systems, such as airbag, ABS and ESP, is so far to a large extent based on functional structures and layers in the nanometer range.

Figure 6.6 LED tail lights are state-of-the-art in automobiles.

Electronics

Recently, electronics has stirred up a real "innovation avalanche" in car manufacture. The steadily growing number of electronic systems in automobiles and their increasing network through bus systems will require electronic state-of-the-art systems in future, which are consequently nanotechnological products. Current processors or chips contain functional structures in the range of 90–130 nanometers; in a few years, this size will shrink to 65 and then to 45 nm. This will be accompanied by a giant leap upwards in the efficiency of electronics and a giant leap downwards in costs, as it is necessary for future car systems. With the help of competitive nanoelectronics, passive safety systems, such as ABS and ESP, will develop into active safety systems that recognize dangerous situations in good time and are able to avoid obstacles actively.

In a workshop, representatives of German companies from the fields of car manufacture and suppliers were questioned about the priorities of economic relevance of different nanotechnological developments until the year 2006. The results are summarized in Fig. 6.7.

The highest relevance for the economic potential of nanotechnological applications in car manufacture until the year 2006 was attributed to:

Components/systems:
- Varnishes (e.g. scratch-resistant varnishes, effect varnishes)
- Outer skin/surface functionalization (anti-reflection layers, anti-fog layers, easy-to-clean layers etc.)
- Tires (nanoscale fillers)

Fields of application
- Light-weight manufacture (nanocomposites, functional layers etc.)
- Wear reduction (tribological layers, nanomaterials etc.)
- Emission control (catalytic converters, filters, fuel additives etc.)

In the following, the value-added chains of nanotechnology in car manufacture are explained in detail for some examples evaluated as economically relevant in the workshop. Table 6.5 summarizes information on applied nanomaterials, intermediate products, applications, technologies and on the product features improved by nanotechnology (by their nanotechnological relevance) in car manufacture.

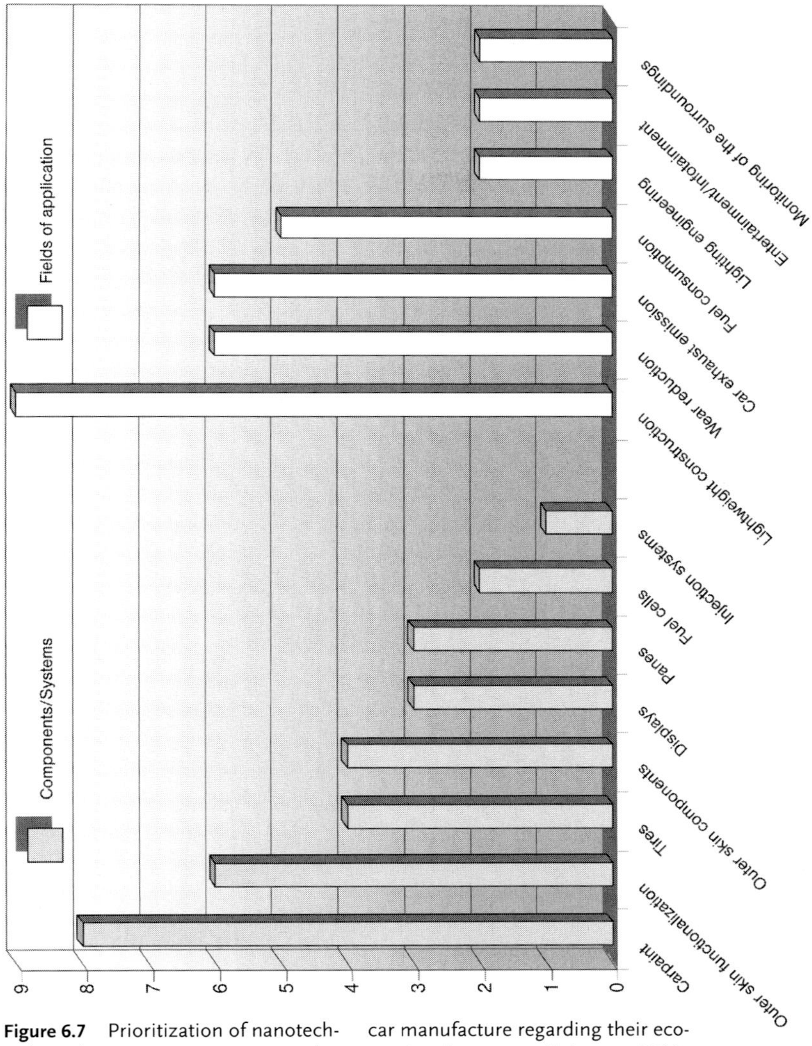

Figure 6.7 Prioritization of nanotechnological components/systems and their fields of application in the field of car manufacture regarding their economic relevance until the year 2006.

Table 6.5 Description of value-added chains for different prioritized nanotechnological applications in car manufacture.

Tools/nanomaterials	Components/systems	Applications	Nanotechnological relevance
• Steels • Polymer nanocomposites • Light metals • Gradient layers • Ceramics	• Bodywork • Supporting structures • Drive • Polymer plates	Light-weight manufacture for improved fuel efficiency	Nanoscale structures, layers improve material properties: • Higher stability • Tribology • Scratch-resistant polymer plates
• Lubricants • Nanoparticles • Nanoporous catalyst support • Membranes	• Catalytic converters • Fuel additives • Particle filters • Alternative drives	Reduction of exhaust emission	Increase in efficiency through higher catalytic reactivity (surface–volume ratio, selectiveness etc.)
• Lacquer additives (UV-protection etc.) • Pigments • Paintwork process	• Effect varnish • Scratch-resistant clear varnish	Car paintwork, color effects, scratch resistance	Nanoscale basic materials result in improved product features
• Raw materials • Fillers • Additives • Nanoanalytics	• Car tires	Less rolling resistance, higher durability and grip	Nanoscale structures, layers improve material properties

6.4.2
Market Potentials

The evaluation of the market potential of nanotechnology in car manufacture can only be quantified with difficulty. Although nanotechnologically influenced components are already a standard feature in car manufacture, the share of nanotechnology in the value added is not usually identifiable, since the application of nanotechnology often comprises only a process step in the manufacturing of complex components, for example, the coating of headlight reflectors, diesel-injectors etc. Even on the level of key elements and materials, concrete evaluations of the market potential of nanotechnology in car manufacture are, at best, possible for certain sections, since as a rule, an industry-specific segmentation of sales figures of nanomaterials is not available. For example, only a part of the world sales of nanomaterials, such as carbon black, to the amount of approx. 3 billion EUR per year, is reached for applications in car tires. Within the context of the company survey and the expert interviews, only a few quantitative market

Table 6.6 Assessments of the world market of nanotechnological products in car manufacture. Source: [1] company survey, [2] expert interviews.

Nanotechnology components/systems	Annual world market volume (reference year)
magnetoelectronic sensors in automobiles	600 million USD (2006)[2]
antifog-coatings of headlight coverings	50–250 million EUR (2003)[1]
varnish	50–250 million EUR (2006)[1]
car tires	7 billion EUR (2006)[1]
components with nanoscale hard coatings	0.5–1 billion EUR (2006)[1]

assessments for applications of nanotechnology in car manufacture could be gathered, although the value-added share of nanotechnology in the components mentioned is not quantifiable (see Table 6.6).

Within the context of an expert workshop, representatives of German nanotechnology and automobile companies regarded the short-term market relevance of nanotechnology in car manufacture as relatively low. The reasons for this are the rather long lead times for technological developments, usually dependent on the innovation cycles of different model lines. Moreover, high quality and safety standards in car manufacture complicate a short-term realization of technological innovations, as usually lengthy test procedures have to be gone through. A moderate market growth until 2006 is expected in the fields of polymer-nanocomposites, nanovarnishes and nanocoated

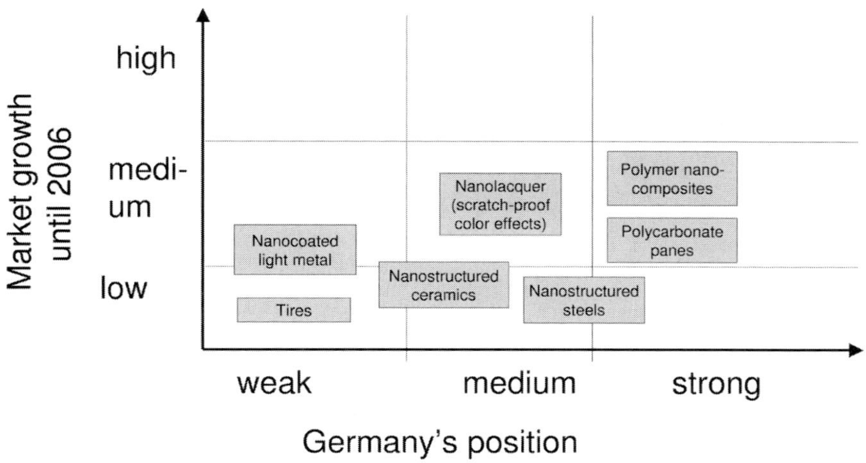

Figure 6.8 Evaluation of prioritized nanotechnological applications in car manufacture, with reference to Germany's position and the market growth until 2006 (Source: expert workshop).

car windows, in which German companies have a good starting position as well.

In the long run (from 2010), however, a considerably increasing influence of nanotechnology in car manufacture is expected, with the corresponding effects on the market potentials especially in the supply industry of car components. Here, a dynamic sector will be automobile sensor technology. Magnetoelectronic sensors based on GMR (giant-magnetoresistive) respectively TMR (tunnel-magnetoresistive) sensor layer systems are expected to gain significant market shares in the field of pressure/expansion measurements, for instance for tire pressure control/drivability systems. The world market volume for magnetoelectronic sensors in car manufacture is assessed at about 600 million EUR in the year 2006, with German component suppliers being in a good starting position (source: expert interviews).

6.5 Optics

The optical industry is usually subsumed under the industry cluster "precision engineering and optics", which also comprises mechatronics and medical technologies apart from optical technologies.

These are fast-growing lines of business, high-technology products of which are applied in nearly all branches of industry, such as in the information and communication technology, in computers, in car manufacture, in photography or in non-contact measuring technology. As key and cross-sectional technologies they are of great importance for the industrial location of Germany. Precision engineering and optics are primarily medium-sized lines of business. Far more than half of all German inventions come from small or medium-sized enterprises, and these special solutions are often the forte of the small companies. The optical industry uses to a large extent most modern nanotechnological manufacturing methods and equipment in their application fields of consumer optics (ophthalmic optics), photonics and precision technology (imaging/photo technology, laser/optical components) as well as in measuring and sensor technology. The BMBF has realized the importance of optical technologies, and with the German agenda "Optical Technologies for the 21st Century", it gives the signal to set off into the future era of the photon. The optical industry is a very research-intensive line of business. On average, good 9% of the turnover – in individual cases up to 25% – is spent on research and development (R&D), with an upward trend. In addition, with an average of approx. 70%, the share of skilled workers is extremely high. The optical and precision-mechanical industry employed about 172 000 people in the year 2002, 21 000 of which in the field of consumer optics, about 14 000 in the field of imaging and photo technology, approx. 64 000 in the field of measuring and sensor technology and about 40 000 in the field of laser and optical components (Spectaris, 2003).

6.5.1 Value-added Chains and Application Potentials

The value-added chains in the field of nanooptics are to be subdivided roughly into the fields of:

- Key materials/procedures
- Nano-optical intermediate products
- Final products in the different fields of application

Fig. 6.9 gives a rough overview of the key materials/procedures, components and systems relevant in the field of nanooptics.

In the following, economically relevant fields of application and product groups of nanooptics are summarized once again (see also Chapter 3).

Ultraprecision optics

Ultraprecision optics is mainly applied in lithography that requires atomically precise optics for the manufacturing of electronic components with nanoscale structural sizes, especially at the transition to the EUV-range.

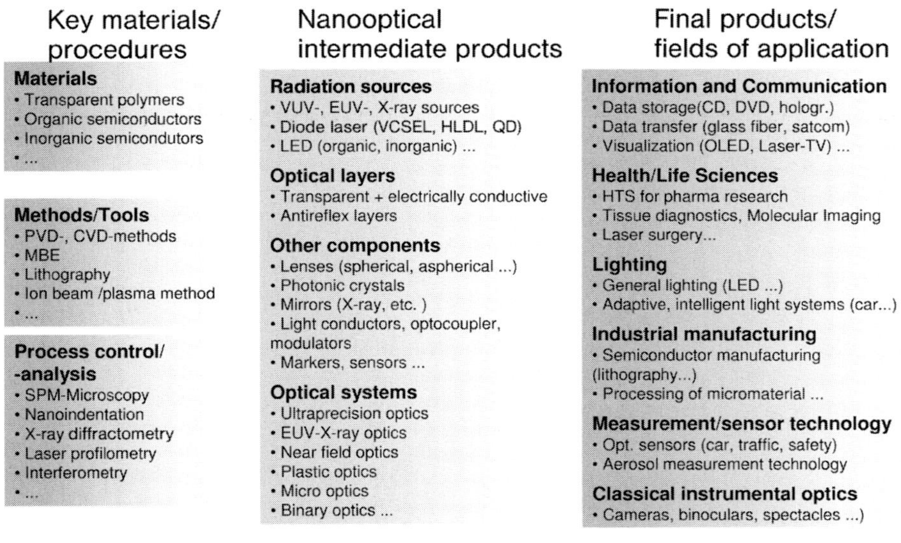

Figure 6.9 Key materials/procedures, components and systems in the field of nanooptics.

Optoelectronic Light Sources – Lasers and Light-emitting Diodes (LED)

Optoelectronic components such as laser diodes and LED are based on extremely thin, only nanometer thick semiconductor layers. Such components have been used in high-volume mass markets for a long time, especially in the fields of I&C technology (e.g. diode lasers for DVD and CD equipment), lighting engineering (LED) and other fields of application. Further developments of the years to come are aimed at new wave-length ranges, the improvement of light power, efficiency and durability as well as the development of flexible light sources on polymer basis. With this, it will probably be possible to open up further promising markets, for example laser-TV or even white LED as headlights in automobiles.

Figure 6.10 White LED exhibit potential as energy-efficient light sources in lighting engineering.

Flat displays

In future, flat screens will largely replace conventional cathode-ray tubes and may achieve a market volume of 64 billion USD in the year 2006 (Becker, 2003). The main share goes to liquid crystal displays, but also other concepts based on nanotechnology, such as OLED or FED, will gain significant market shares. OLED displays are mainly expected to lead to large cost-savings and new possibilities of applications as they allow large-area processing and flexible and economical manufacturing. The principle functioning of an OLED (organic LED) is based on injection electro luminescence, similar to inorganic LED, in which nanoscale layers of organic semiconductors play a function-determining role. FED are self-luminous and can depict colorful pictures with high brightness and sufficient contrast. With this technology, highest resolutions can be achieved and the development of large-area displays is possible without fundamental problems. Production may occur with the help of thin-film technology which allows economic mass production. In modern FED, carbon nanotubes are used as particularly efficient electron emitters (CNT-FED).

Figure 6.11 Mechanically flexible displays of self-luminous organic LEDs (OLED) (source: Siemens).

Optical measurement and sensor technology

Optical sensors in the visual, infrared and ultraviolet spectral region have an increasing penetration in different industrial fields of application. Also in this field, nanotechnological applications may improve efficiency, selectiveness and durability of sensor components.

In a workshop, representatives of German companies active in the field of nanooptics were questioned about the priorities of the economic relevance of different nanotechnological developments in the range of intermediate products (components/systems) and application fields until the year 2006. The results are summarized in the Figs. 6.12 and 6.13.

German nanooptics companies considered the fields of laser diodes and LEDs, functional layers and ultraprecision optics (especially lithographic systems for the manufacturing of semiconductors) to be economically important. The value-added chains for the given examples of nanooptics are summarized in Table 6.7 and characterized according to the applied nanomaterials, intermediate products, applications, technologies and product features (respectively their nanotechnological relevance) improved by nanotechnology.

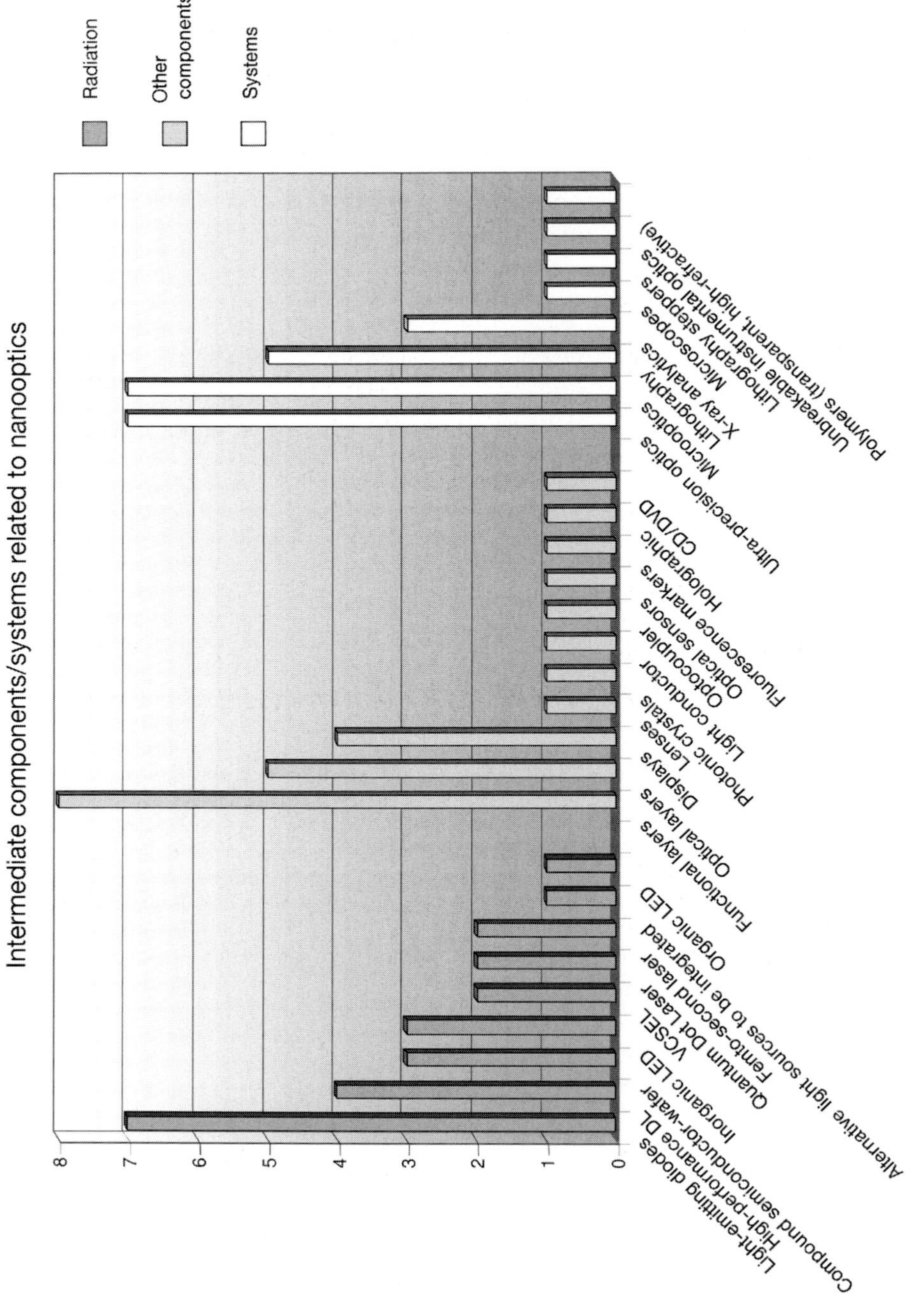

Figure 6.12 Prioritization of components/systems in the field of nanooptics with regard to the economic relevance until the year 2006.

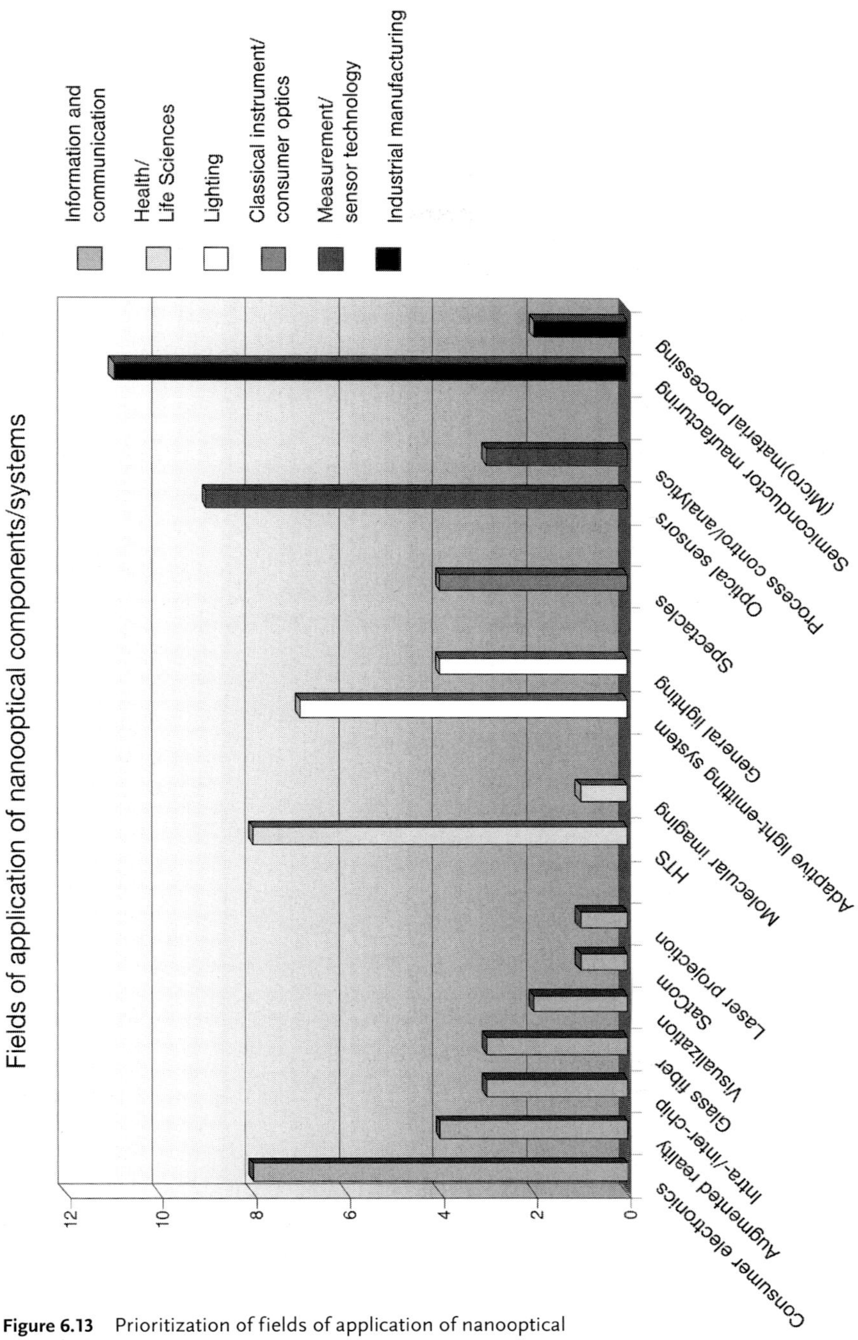

Figure 6.13 Prioritization of fields of application of nanooptical components/systems with regard to the economic relevance until the year 2006

Table 6.7 Description of the value-added chains for different prioritized nanotechnological applications in the optical industry.

Tools/materials	Components/systems	Applications	Nanotechnological relevance
• Design and simulation • Ultraprecision optics • Nanostructuring • Coatings	• Light sources • Masks • Imaging optics	Semiconductor manufacturing: • Optical lithographic systems	• Tools for nano-structuring • Ultraprecise surfaces • Freedom from defects (masks)
• Ion beam/ sputter process • Spin coating • CVD process • Sol–gel	Functional layers • Mech. protection • Scratch resistance • Phosphorus layers • Opt. coupling layers	• Medical endoscopy • Biosensoring technology • Displays • Lighting • Glasses	• Optical properties in comb. with other functionality • Interface engineering • Unique selling proposition
• MBE, CVD, PVD • Optical lithography • Thermal processes	• LEDs • Laser diodes	• Opt. data transfer and storage • Lighting • Medicine • Measuring technology + analytics	• Defect-free layers, • Nanostructures (quantum dot)
• Opt. lithography • Coating • Laser micromaterial processing	• Diffractive optics • Lenses/mirrors • Bragg grid	• Holography • Spectroscopy • Microoptics • Optical data networks • Displays	• Geometrical requirements

6.5.2
Market Potentials

Market potentials resulting from nanotechnology in the optical industry arise mainly in the field of ultraprecision optics (especially in the semiconductor manufacturing) and in the field of optoelectronic light sources (laser diodes and LED). In Table 6.8, the assessments of the market potentials for different nanotechnological components and systems are compiled. Here it has to be taken into account that, as a rule, the share of nanotechnology in the value added of the products mentioned is not quantifiable. However, nanotechnology contributes essential functionalities that are indispensable for the competitiveness of the products and their success on the market.

As far as market growth up to 2006 is concerned, the product group of diode lasers and LED is regarded as the most dynamic field. For the field of LED, annual growth rates of 80% are expected, for instance.

Table 6.8 Assessments of the world market for nanotechnological products in the field of optical industry. Sources: [1] company survey, [2] Fecht et al. 2003, [3] expert workshop.

Nanotechnological components/systems	Annual world market volume (reference year)
Ultraprecision optics, of which lithographic optics	1–5 billion EUR (2006)[1] 0.5–1 billion EUR (2006)[1]
Lithographic steppers	7.7 billion USD (2006)[2]
LED, of which white LED	1–5 billion EUR (2006)[3] 10–50 million EUR (2006)[3]
Diode lasers, of which high-performance diode lasers	1–5 billion EUR (2006)[3] 50–250 million EUR (2006)[1]
OLED displays	0.1 billion USD (2002), 2.5 billion USD (2006)[2]
CNT-FED	0.01 billion USD (2002), 0.05 billion USD (2006)[2]
Optical sensor technology	1–5 billion EUR (2006)[3]
Laser interferometer	10–50 million EUR (2006)[1]
Optical thin-film measuring technology	250–500 million EUR (2006)[1]

According to the evaluation of the German company representatives, Germany is well positioned in these fields and holds a leading role. Regarding the commercial realization, especially for the end-user market (e.g. displays), Germany's position in contrast to Japan and Southeast Asia is considered to be significantly weaker (source: expert interview). In addition, Germany is said to have a strong position in the fields of optical lithography, ultraprecision optics and functional

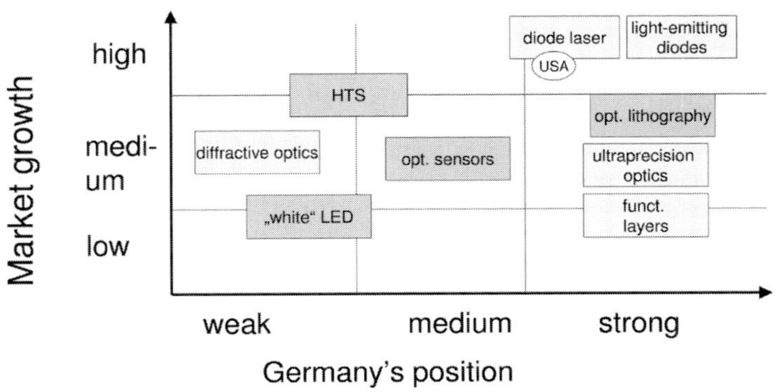

Figure 6.14 Market growth and Germany's position in the field of nanotechnological products in the optical industry.

coatings. In the dynamic market of high-throughput screenings, however, in which nano-optical detection systems play an increasingly important role, Germany is only poorly represented.

6.6 Medicine and Life Sciences

Use of nanomaterials in life sciences will have far-reaching effects on our social life

The field of life sciences comprises a group of natural scientific branches of research dealing with the specific free-market application of scientific insights of modern biology, biotechnology and other fields. These insights contribute decisively to the understanding of living organisms and ecological systems. This opens up possibilities undreamt of in the explanation of genetically caused illnesses or of those triggered by external influences, and it enables the development of new therapies. The fundamental new insights into this scientific field, progress in the related technologies and their broad application in health, environment and food research will have far-reaching effects on health care, the control of environmental dangers and our entire social life. At the same time, life sciences offer a great potential for the creation and the maintenance of new and high-quality jobs.[7] Life sciences are mainly relevant for the following branches of industry:

- Medicine/pharmacy
- Food industry
- Cosmetics

6.6.1 Value-added Chains and Application Potentials

6.6.1.1 Medicine/pharmacy

Intensive research activities in the fields of drug delivery and new kinds of biochips

In the field of nanotechnological applications for medicine and pharmacy, intensive research activities in the fields of drug delivery and new kinds of biochip systems has taken place recently (see Wagner and Wechsler, 2004). Moreover, there are a lot of approaches to apply nanoparticles in the molecular diagnostics and to use nanostructured materials for the development of bioactive surfaces. The field of life sciences was the focus of a detailed parallel study [8] on applications of nanotechnology, so only a summary of the essential fields of application is given here.

7) see www.bmbf.de

8) As consortium leader, the Aachener Gesellschaft für Innovation und Technologietransfer (AGIT) has drawn up a study on nanotechnology and health published in autumn 2004.

Biomedical basic research

The continuous development of instrumental analysis, such as scanning probe techniques or optical single-molecule spectroscopy, allows the examination of biological objects on the nanoscale. This contributes much to the understanding of biochemical processes and in the long run, it may lead to new strategies for the control of illnesses.

Drug delivery

Nanoscale drug delivery systems offer the potential to deliver poorly soluble or chemically unstable active substances to the damaged tissue in aqueous media, to overcome biological barriers such as the blood–brain barrier and to accumulate the active substances directly in the damaged tissue in order to diminish the danger of side effects. The first drugs using such drug-delivery systems are already on the market.

First nanoscale drug delivery systems already on the market

Contrast media in diagnostics

Nanoparticles can be used for molecular imaging to accumulate contrast media in damaged tissue using molecular markers. Since the molecular signatures of many diseases appear even before the symptoms break out, these methods enable the diagnosis of diseases in the early stages.

Biochips

Another field where nanotechnology is gaining importance is biochip technology. On a long and medium-term basis, such compact systems will open up a mass market in the medical diagnostics, especially to replace expensive and time-consuming laboratory tests by quick tests on the spot with the medical practitioner or in hospital. Such systems are also being developed for application in personalized medicine. In this field, nanotechnology will not only play a role in further miniaturization, but also for the improvement of detection sensitivity and reliability of the systems For example, nanoparticle fluorescence marker or near-field optical detection systems are used to increase the detection sensitivity of biochips. Apart from this, a lot of new types of electric and magnetic biochip detection systems based on nanotechnology are currently being developed. Exactly in this field, German research institutes and companies are very active and internationally well-positioned. In comparison with conventional optical processes, electric and magnetic detectors are more robust and easier to integrate into a miniaturized sensor.

Implants

With nanostructured surfaces of implants, their biocompatibility and ingrowth can be improved. Great importance is attached to the better understanding of the processes at the interface between the tissue and the surface of the implant.

6.6.1.2 Cosmetics

Cosmetics will profit mainly from the application of nanomaterials. Among other things, nanoparticles are already used in the following sectors:

Sun-protection products

Nanoparticles as UV-absorbers in sun-protection products

Nanoscale metal oxide particles such as titanium dioxide or zinc oxide are already used as highly efficient UV-absorbers in sun protection products. Owing to their small particle size, the dispersions used are transparent and offer sun protection invisible on the skin. Another advantage of inorganic UV-absorbers compared to organic ones is the lower allergenic potential. However, it is not definitely clarified whether, under certain circumstances, nanoparticles might reach the human organism through the skin and cause systemic effects there. The world market for nanoparticles used in sun protection is assessed at 87 million USD for the year 2005 (BCC, 2003).

Pigments

Nanostructured pigments are used in cosmetic colors in order to obtain special color effects. Among other things, such pigments are based on nanoscale coated silicate plates that, depending on the perspective, change color owing to interference effects.

Emulsions

Nanoscale hollow bodies, such as liposomes, are used as transport containers for different active substances in skin creams. While liposomes mainly transport active substances soluble in water, nanosomes are rather used as vehicles for fat-soluble substances, such as vitamin A or E; nanosomes consist of a lecithin cover, enclosing an oil-containing nucleus.

Toothpaste

Nanoparticles as additives in toothpaste for the regeneration of defective tooth enamel

Nanoparticles are currently being developed as additives in toothpaste for the regeneration of defective tooth enamel. SusTech, for instance, is developing an active substance of calcium phosphate protein particles that are able to penetrate the dentinal canal of the teeth easily. There, they serve as crystallization germs which take up further minerals and close the canals, thus defusing the problem of sensitive, exposed necks of the teeth.

6.6.1.3 Food Technology

Sectors in food technology with application potential for nanotechnology are for example:

Packaging material

Nanotechnology has already been used in the field of food packaging materials. Nanoparticle-reinforced polymers that have low gas permeability and are therefore suitable for the application as food films. Coating technologies are also applied to enhance the gas tightness of PET-bottles for beverages, for instance. In this case, both nanoscale inside and outside coatings are used that are produced by plasma processes.

Food additives

Through nanostructuring of special active substances, such as vitamins used as food additives, their effectiveness in the human organism will be enhanced. Micronized active substances have already been used to a large extent and achieve a world market volume of approx. 1 billion USD (BASF, 2002). Nanoparticles are already used in food to achieve color effects or to improve, for example, the flowing properties of ketchup.

Biosensoring technology

In the food sector, biosensors able to control the freshness of food are of interest. Thin-film sensors are being developed for the identification of volatile compounds that could be integrated directly into the packaging and could indicate the spoilage of food, for example, by color alteration.

6.6.2 Market Potentials

Up to now, only a few market evaluations for applications of nanotechnology in the sector of life science have been available. An overview of existing market forecasts is given in Table 6.9.

Table 6.9 Evaluations of the world market of nanotechnologically influenced products in the field of life sciences.

	World market volume (year)	Source
Medicine/pharmacy		
Biophysical analytics (e.g. scanning probe techniques)	181 million USD (2002) 745 million USD (2007)	BCC 2003
Total market biochips/quick tests	2 billion USD (2010)	VDI News 2003
DNA chips	1.9 billion USD (2006)	DB 2003
Protein chips	0.4 billion USD (2006)	DB 2003
Nanotechnology-based diagnostics and analytics (e.g. nanoparticles for biochips, biomagnetic separation, contrast media)	80 million USD (2002) 391 million USD (2007)	BCC 2003
Active substances and drug delivery	8 million USD (2002) 33 million USD (2007)	BCC 2003
Tissue engineering	0 million USD (2002) 1.5 million USD (2007)	BCC 2003
Ag-nanoparticles in antimicrobial applications	1 million USD (2005)	BCC 2001
Cosmetics		
Nanoparticles in sun protection products	86.50 million USD (2005)	BCC 2001

6.7 Electronics

Electronics is a key technology

Semiconductor electronics has an important leverage effect on the electronics market. Electronics in turn is a key technology itself for different lines of industry. For the production of semiconductor electronics, the additional supply of equipment and materials is required. Figure 6.15 depicts the value-added chain in the form of a chart.

Figure 6.16 compares the development of the market for semiconductor electronics and the related equipment market with the development of the electronics market and the global gross domestic product. The market for semiconductor electronics has been showing an average annual growth of 14.7% since 1965. That is, it has grown twice as fast as the global gross domestic product.

According to a forecast of the World Semiconductor Trade Statistics (WSTS), turnovers of worldwide 215 billion USD will be achieved in the production of semiconductor electronics in the year 2006 (Fig. 6.17). The vagueness of such forecasts is relatively high due to wild market fluctuations.[9] So BCC forecasts sales of 246 billion USD for

9) Since 1965, the annual growth rates have ranged between −32.0% and +44.6%.

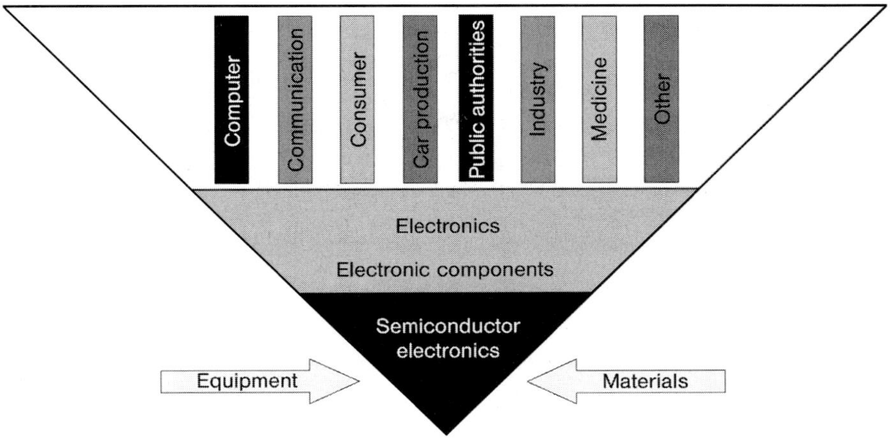

Figure 6.15 Increase in value in electronics.

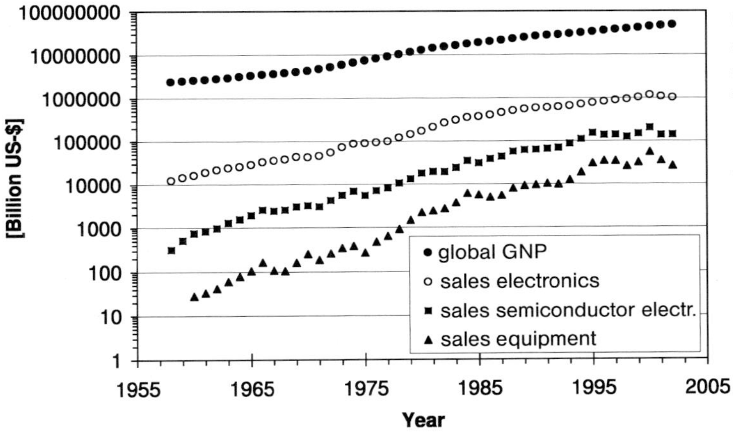

Figure 6.16 Historical development of the market for semiconductor electronics and the equipment market in comparison with the electronics market and the global gross domestic product (Source: VLSI Research, 2004).

2006, while the Semiconductor Industry Association (SIA) goes on the assumption of 205 billion USD. All in all, growth above average is expected in the economic area of Asia /Pacific.

Figure 6.18 shows the division of the production of semiconductor electronics into different components. According to this forecast of the SIA, the logic sector has the largest share and will put on the sharpest growth in the years to come.

The market for equipment related to the production of semiconductor electronics recorded an even sharper growth than the market for semiconductor electronics; however, it is also subject to wilder economic fluctuations. Fig. 6.19 shows the development of the market according to a forecast of VLSI Research until 2008.

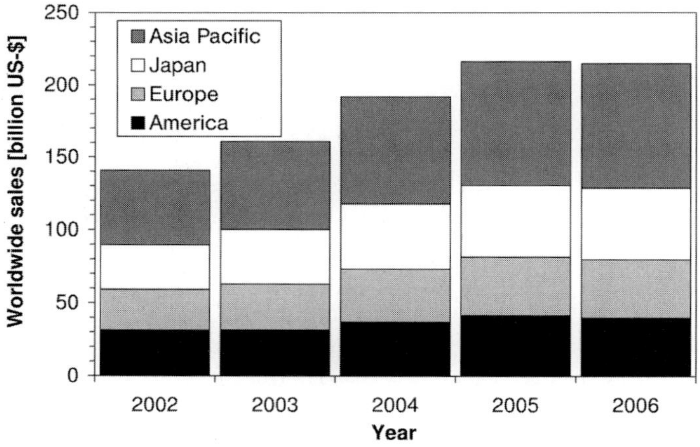

Figure 6.17 Development of the global production of semiconductor electronics (Source: WSTS).

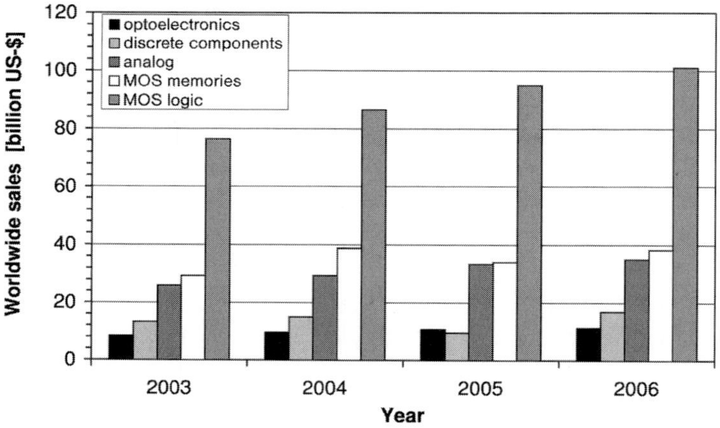

Figure 6.18 Development of the market for semiconductor electronics, subdivided according to different components (source: SIA, 2004).

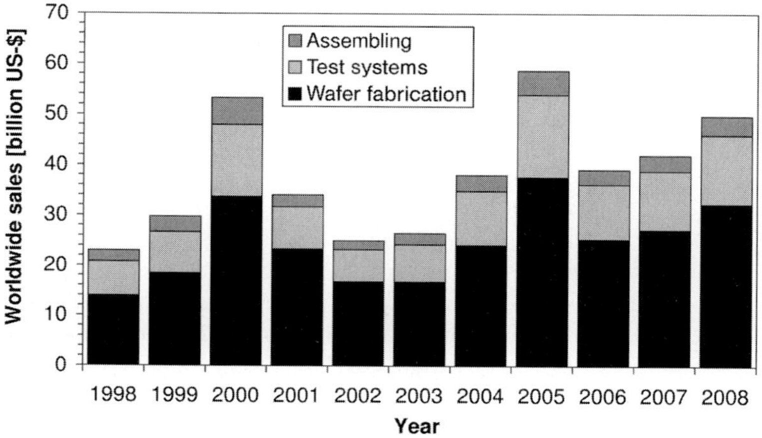

Figure 6.19 Development of the worldwide sales of equipment for the production of semiconductor electronics until 2008 (source: VLSI Research, 2004).

According to data of the market research company Gartner, 21 billion USD were turned over with materials for the production of semiconductor electronics in 2002.

The Zentralverband Elektrotechnik- und Elektroindustrie (the umbrella organization for electrical engineering and electrical industry, ZVEI) assesses the market for electronic components at worldwide 300.5 billion USD for 2003. 17.23 billion USD of it are attributable to the German market and 10.59 billion USD of it can be assigned to the semiconductor-electronics sector. The most important buyer of electronic components in Germany is the car industry with a share of nearly 32%, followed by data engineering with 24%, telecommunications with 22% and the industrial electronics with 17%. With a market share of 5%, consumer electronics, which is only of secondary importance in Germany, brings up the rear of the list (source ZVEI, VDI news, 21.11.2003).

Since 1965, the whole market for electronics shows an average annual growth of 10.3%. According to the market research institute, VLSI Research, 1.168 billion USD were turned over with electronic products worldwide in 2004. For the year 2006, sales of 1.329 billion USD are forecasted. Figure 6.20 shows the division of the market for electronics into different lines of industry. It is noticeable that there will be only slight shifts of the market shares until 2008.

In Europe, Germany, France and Great Britain are leading the electronic production, as depicted for the year 2001 in Fig. 6.21. More detailed and more current figures regarding the information and communications market (I&C market) come from the European Information Technology Observatory (EITO). For 2004, the global I&C market

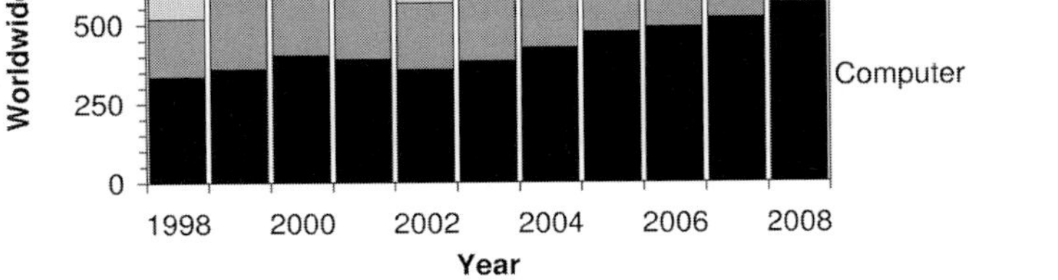

Figure 6.20 Development of the global electronic production until 2008, subdivided according to industries (source: VLSI Research, 2004).

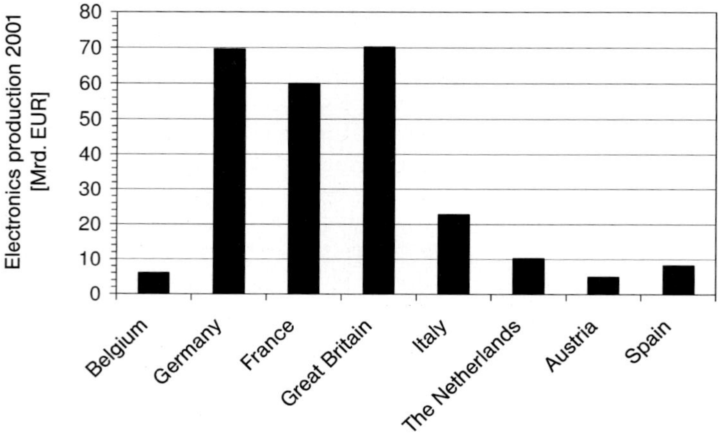

Figure 6.21 Electronic production in Europe. Data of the Association European Electronic Components Industry is given for the year 2001.

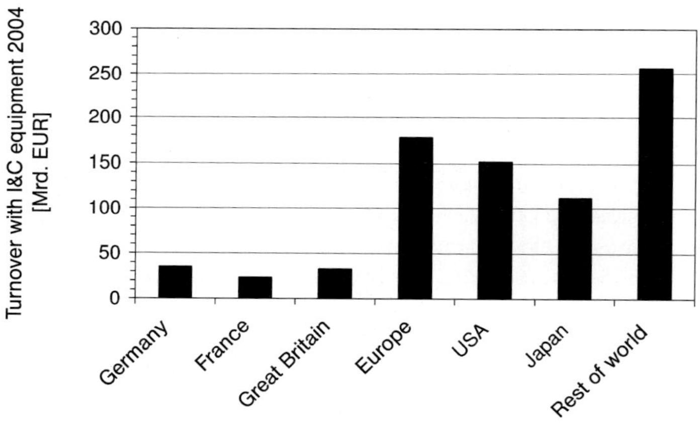

Figure 6.22 Turnover with I&C equipment in the year 2004 (source: EITO, 2003).

is assessed at a volume of 2378 billion EUR. The electronic share is hidden behind the sector I&C-equipment[10)] with a volume of 696 billion EUR. Here again, Germany is the leading country in Europe (see Fig. 6.21).

Over recent decades, the share of semiconductor-electronics in electronics grew from approx. 4.5% in the year 1960 to currently approx.

10) In this category, the EITO subsumes the fields of computer hardware, office equipment, end user communications equipment and datacom and network equipment.

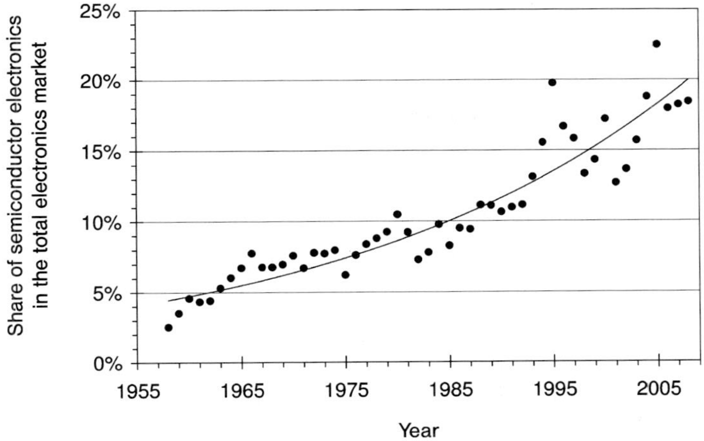

Figure 6.23 Market share for semiconductor electronics in the entire electronics market (Source: VLSI Research).

17.5% and will continue to grow (see Fig. 6.23). According to the pan-European program MEDEA+, the share of semiconductor electronics in mobile phones and other high-tech-applications will be at approx. 50% in 2007.

As already mentioned in Chapter 3, the market for semiconductor electronics will still be dominated by silicon-based CMOS-electronics. In contrast to other technologies and driven by continuous miniaturization, a continuous transition from microelectronics to nanoelectronics clearly predetermined by the ITRS will take place.

Until now, figures indicating the share of the 100-nm-generation or of following generations in the total production of semiconductor electronics have not been available. Figure 6.24 shows how the wafer-starts per week are shared between the production lines larger than 120 nm and smaller than 120 nm. According to this, the share of the generations smaller than 120 nm will grow continuously reaching a share of 12.5% until 2006.

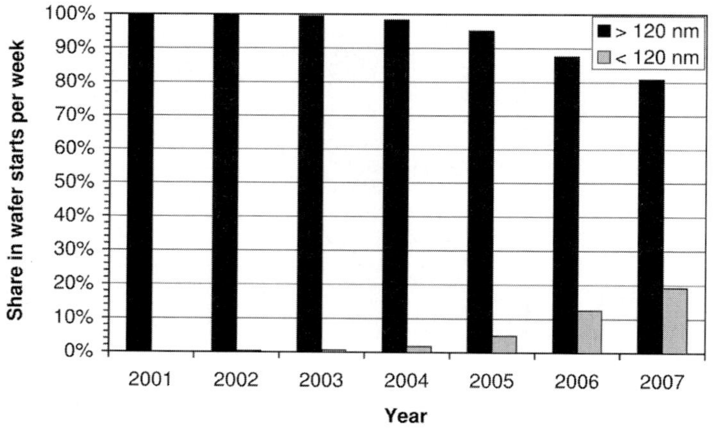

Figure 6.24 Shareout of wafer starts per week to the CMOS-generations larger than 120 nm and smaller than 120 nm (source: VLSI research).

Market for nanoelectronics 2006: more than 20 billion USD

These figures allow the assessment that nanoelectronics will make up approx. 10% of the market for semiconductor electronics in the year 2006. According to the forecasts, this corresponds to a market volume of more than 20 billion USD.

Figure 6.25 gives an overview of chip manufacturers, suppliers and system manufacturers domiciled in Germany. According to the Association for Information Technology, Telecommunications and New Media (Bitkom), the number of jobs in the production of electronic components amounted to 82 000 in 1999. Afterwards, the Association did not publish relevant figures again. A systematic survey with regard to jobs in the fields of semiconductor electronics, equipment and electronics has not been carried out yet. According to assessments of the BMBF, there are altogether 215 000 jobs in this field, 20 000 of them in the chip manufacturing, 45 000 at suppliers of equipment and 150 000 at system manufacturers.

Figure 6.25 Locations of the electronics industry in Germany (incomplete) (source: BMBF).

6.8 Overview of Market Potentials and Time Horizons

Table 6.10 summarizes the market potentials of nanotechnological applications in the Lead Markets described. However, it is not possible to evaluate the "world market for nanotechnology" on the basis of the figures given in publicly available studies, since:

- market information is only available for part of the nanotechnological products and lists are therefore incomplete;
- the market forecasts partly refer to different time horizons;
- nanotechnological products are named twice in two or more sections (e.g. application of basic nanoproducts/components in end products of different branches of industry);
- the survey includes products from different stages of the value-added chain (basic products, intermediate products, end products etc.).

Table 6.10 Assessments of the annual world market volume of nanotechnological products in the respective Lead Markets.

Nanotechnological products	Annual world market volume (reference year)
Chemistry	
Nanomaterials	
• Metal oxide/metal nanoparticles	900 million USD (2005)[1]
• Nansilicic acid	800 million EUR (2003)[2]
• Nanolayer silicates	25 million USD (2006)[3]
• Carbon nanotubes	145 million USD (2005)[4] 1.2 billion USD (2006)[5]
• Carbon black	3 billion USD (2002)[6] 8 billion USD (2006)[5]
• Polymer dispersions	15 billion EUR (2002)[7]
• Organic semiconductors	500 million. USD (2005)[10]
• Dendrimers	5–15 million EUR (2006)[3]
• Micronized active substances	1 billion EUR (2002)[7]
• Zeolites	2.6 billion USD (2006)[5]
• Aerogels	10 billion USD (2005)[8]
• Polymer nanocomposites	0.3 billion USD (2006)[3] 1.5 billion EUR (2009)[9]
Intermediate products	
• Corrosion protection paper	10–50 million EUR (2006)[12]
• Varnishes	50–250 million EUR (2006)[12]
• Films for displays	50–250 million EUR (2006)[12]
• Marker substances	250–500 million EUR (2006)[12]
Nanosensors	
• Temperature sensors	4.6 million USD (2004) 217 million USD (2011)[11]
• Pressure sensors	4.4 million USD (2004) 87 million USD (2011)[11]
• Chemical sensors	1.3 million USD (2007) 36 million USD (2011)[11]
Car manufacture	
• Magnetoelectronic sensors	600 million USD (2006)[13]
• Antifog-coatings for headlights	50–250 million EUR (2003)[12]
• Varnishes	50–250 million EUR (2006)[12]
• Car tires	7 billion EUR (2006)[12]
• Components with hard coatings	0.5 –1 billion EUR (2006)[12]
Electronics	
• CMOS-electronics <100 nm	20 billion USD (2006)[18]
• GMR-HDD	26.6 billion USD (2006)[5]
• MRAM	30–50 billion USD (2010)[19] (DRAM–replacement)

Table 6.10 Continued

Nanotechnological products	Annual world market volume (reference year)
Optical industry	
• Ultraprecision optics,	1–5 billion EUR (2006)[12]
thereof lithography optics	0.5–1 billion EUR (2006)[12]
• Lithography-steppers	7.7 billion USD (2006)[5]
• LED,	1–5 billion EUR (2006)[13]
thereof white LED	10–50 million EUR (2006)[13]
• Diode lasers,	1–5 billion EUR (2006)[13]
thereof high-efficiency diode lasers	50–250 million EUR (2006)[12]
• OLED-displays	0.1 billion USD (2002) 2.5 billion USD (2006)[5]
• CNT-FED	0.01 billion USD (2002) 0.05 billion USD (2006)[5]
• Optical sensor technology	1–5 billion EUR (2006)[13]
• Laser interferometer	10–50 million EUR (2006)[12]
• Optical thin-film measurement technology	250–500 million EUR (2006)[12]
Life sciences	
Medicine/pharmacy	
• Biophysical analytics	181 million USD (2002) 745 million USD (2007)[14]
• Total market biochips/quick tests	2 billion USD (2010)[15]
• DNA-chips	1.9 billion USD (2006)[5]
• Protein chips	0.4 billion USD (2006)[5]
• Nanobased diagnostics and analytics	80 million USD (2002) 391 million USD (2007)[14]
• Active substances and drug delivery	8 million USD (2002) 33 million USD (2007)[14]
• Tissue engineering	0 million USD (2002) 1.5 million USD (2007)[14]
• Ag-nanoparticles in antimicrobials	1 million USD (2005)[16]
Cosmetics	
• Nanoparticles in sun protection products	86.5 million USD (2005)[16]

Sources: **1** BCC, 2002, **2** Wacker Silicones, 2003, **3** SRI, 2002, **4** Mitsubishi Research Institute, 2002, **5** Fecht et al., 2003, **6** Reuters, 2002, **7** BASF/Distler, 2002, **8** Aspen Systems, 2001, **9** Stevenson, 2003, **10** Frost&Sullivan, 2002, **11** Frost&Sullivan, 2003, **12** company survey, **13** VDI TZ experts workshop, **14** BCC, 2003, **15** VDI-News, 2002, **16** BCC, 2001, **17** Small Times, 2002, **18** own assessment, **19** Small Times, 2003.

Figure 6.26 gives an overview of the stage of development of different nanotechnological products in the Lead Markets examined.

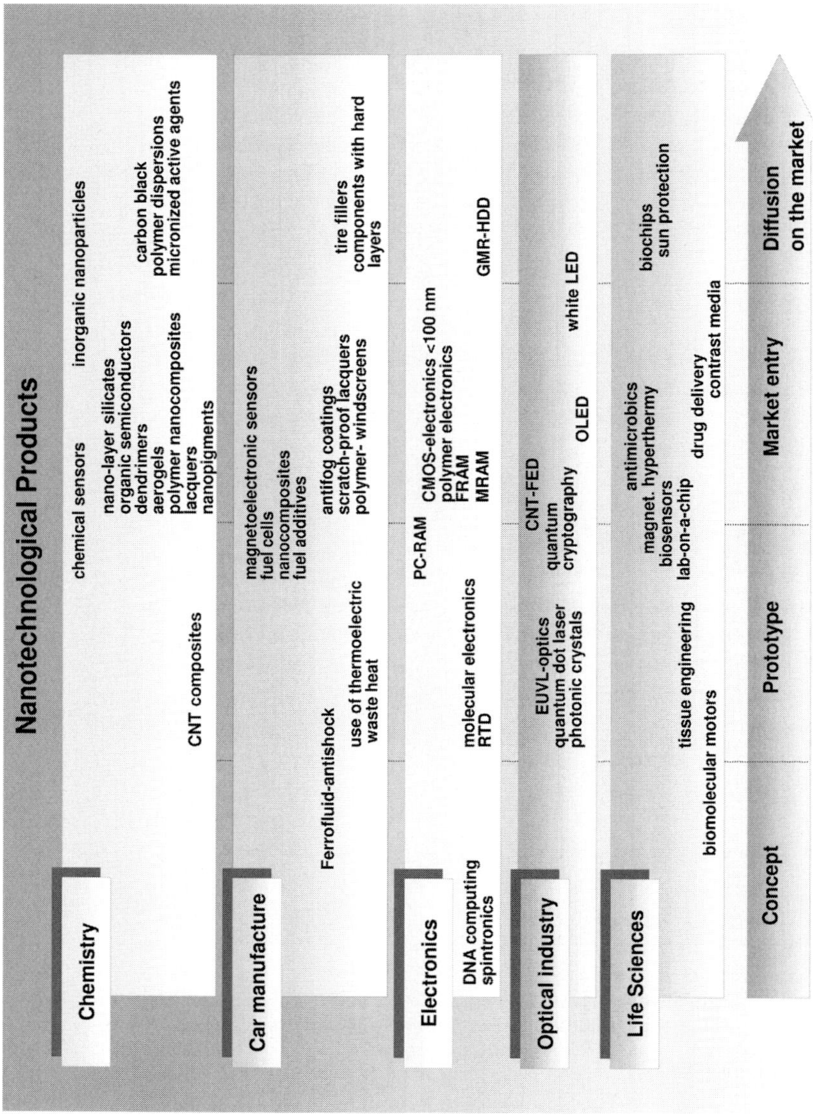

Figure 6.26 Stage of development and time horizons of different nanotechnological products in the examined Lead Markets (incomplete) (source: VDI TZ).

7
Assessment of the Nanotechnological Market in Germany

Matthias Werner, Gerd Bachmann, Norbert Malanowski and Stephan Mietke

7.1
Procedure

Combination and evaluation of total results

This Chapter contains the combination and evaluation of all the results determined. The results are evaluated within the context of results from other commercial sources available (Markus database and the current WMTech market survey, Fecht et al., 2003). This serves to validate the results and to clarify the specific prospects and evaluations for Germany. In this connection, a strength–weakness analysis has been carried out, of the international competitors in this range, of their current and future orientation in the field of research and application, and the comparative position of Germany. Based on the information gathered by structured methods (SWOT analysis and white spot analysis) further analysis of the status in Germany as well as of the existing chances and deficits was carried out.

Position of Germany

A SWOT analysis is a strength–weakness analysis, i.e. evaluation of the factors influenced in Germany itself, and an opportunities–threats analysis, i.e. an assessment of globally acting factors. The white-spot analysis is a qualitative, structured method for the identification of positions weakly occupied or not occupied at all (so-called white spots) by means of a matrix that describes, for instance, markets/applications and technology platforms (Deschamps, 2000). Statements regarding the employment effects of nanotechnology depend on assessments, as is always the case with cross-section technologies. Neither the share of nanotechnology in the gross national product, nor the effects of supporting measures in the field of nanotechnology in the form of created (or axed) jobs can be depicted precisely by statistics. The reasons for this are, on the one hand, the broad diffusion of nanotechnology in a large number of industrial branches in Germany; and on the other hand, the problem of the definition and differentiation of a nanotechnological product, which renders the clear attribution of the industrial output to nanotechnology nearly impossible.

SWOT-analysis

Growth Market Nanotechnology. Edited by N. Malanowski, T. Heimer, W. Luther, and M. Werner

ISBN: 3-527-31457-1

Therefore, within the framework of the employment analysis, it was only possible to assess the direct impact on employment in connection with nanotechnology. For this assessment, the results of the paper-based company survey as well as the available secondary sources, such as the Markus database, company reports and websites have been consulted. The methodology chosen and the limited data availability, mean that it was not possible to analyze the indirect impacts on employment.

Employment analysis

7.2
Nanotechnology Markets for German Companies

7.2.1
Analysis from the Point of View of the Companies Involved

Nanotechnology of great importance in companies

The results of the paper-based survey show (more detailed in Chapter 5) that the companies polled attach great importance to nanotechnology. It becomes clear that the companies in the sample see above all the economic opportunities of nanotechnology and do not regard it as a technological "playground". More than 75% of the companies see that nanotechnology could open up new markets for them. More than 60% of the companies consider nanotechnology to be a decisive competition factor, and/or a chance to improve their technological competitiveness.

In contrast to this, the companies polled disagree with the statement that nanotechnology is only a new field of experiment, just as they reject in a more moderate way the statement that it would extend technological competence (see Figs. 5.21 to 5.24 in Chapter 5). This result clearly refutes the frequent popular opinion that nanotechnology is simply a "hype". Such assertions are partly made by institutions of public finance as well as by the press, usually without reference to competent and profound knowledge or even a qualified or scientific background. However, since it is precisely that funding problem that is one of the most important obstacles to innovation to nanotechnology, cooperation with these fields will be an important challenge for the future; a challenge yet to be addressed especially in Germany.

Companies see considerable sales potential for their nanotechnological products

An essential variable for the companies' evaluation of the future market is their sales assessment. The companies polled have a rather positive attitude towards the potential of their core product in nanotechnology (which represents an extremely broad scope of application for the interviewees as a whole). More than 50% of the companies regard a global sales volume in 2006 of 250 million EUR or more as realistic for their product with the highest priority. A further 26.7% of the companies expect a worldwide market volume of at least

50–250 million EUR (see Fig. 5.37 in Chapter 5). This shows the remarkable sales potentials the companies see for their nanotechnological products.

In this context, the hopes placed on Germany's potential market volume with regard to the total world market are interesting, although a market forecast cannot be derived from this. It becomes clear that in the year 2006, Germany's share in the world market does not correspond with Germany's research potential in nanotechnology at present. While today the research results are considered to be convincing, the market shares range predominantly on average between 11 and 25% (see Fig. 5.38 in Chapter 5). This is a sign of little confidence in the ability of German companies to develop markets.

Convincing research results

The results of the company survey showed that chemistry (including materials) is the most important field of application of nanotechnology companies in Germany, followed by the fields of life sciences (medical technology/health) and I&C. The strength of the chemistry sector in Germany is enhanced by the fact that the majority of the companies questioned see their important competitors in Germany and in the USA.

Strong fields in Germany

In the field of chemistry, the central functional features of nanotechnology are improved material properties and the functionalization of surfaces, followed by protection functions and optical effects, which are absolutely classical and application-related demands.

7.2.2 Market Segmentation

The issue of nanotechnology products is directly related to the problem of how to define a nanotechnology company. A uniform definition does not exist in either case. In contrast to other sectors of technology with relatively homogenous fields of applications – such as biotechnology, for example – in nanotechnology many different companies are active from very different lines of business and with a varying degree of commitment. According to the definition, even heterogeneous products and fields of application are included. As already mentioned, the application fields reach from the manufacturing of nanomaterials over the field of life sciences up to electronics. Following the definition of Ernst & Young which determines three categories of biotechnology companies, a segmentation is used, according to the sales share of nanotechnology as a criterion for its economic importance in the companies. As there is no objective parameter, the statements regarding the assessed share of nanotechnology in the companies' turnover given in the paper-based company survey were used as the basis. 103 companies out of 105 valid questionnaires have been classified according to the results of the paper-based survey. For

What is a nanotechnology company?

the data of the company parameters, only 103 returns have been analyzed, as the other two came from the same company.

In the following, three categories (F1 to F3) are determined that represent a standard for the value added by nanotechnology and for the focus of companies:

Nanotechnology company F1
The sales share in the total sales of the company achieved by nanotechnology products comes to at least 50%.

Nanotechnology company F2
The sales share in the total sales of the company achieved by nanotechnology products ranges between one and 50%.

Company F3
The sales share in the total sales of the company achieved by nanotechnology products is less than 1%. These companies deal with nanotechnology but have not yet reached significant turnovers with it. Here, mainly companies with research activities in nanotechnology are concerned, or those that use nanotechnological products, or even companies in the state of observation with the intention to use nanotechnology in future.

The statistics gathered within the context of this survey among 103 German technology companies (out of 105 utilizable questionnaires) produced the distribution to the three categories F1 to F3, depicted in Fig. 7.1.

The division of companies into three categories enables the development of a nanotechnology indicator that allows the manpower development to be followed chronologically, for instance. Should a comparable parent population in another country, the USA for example, be surveyed a direct "benchmarking" would possible. The lack of historical data regarding the development of sales and manpower as well

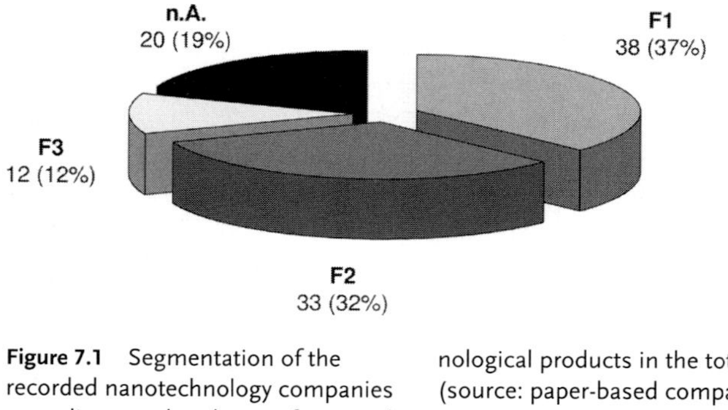

Figure 7.1 Segmentation of the recorded nanotechnology companies according to sales shares of nanotechnological products in the total sales (source: paper-based company survey).

as the corresponding sales shares achieved with nanotechnology, means that the development can be followed only over a very limited period of time. In Section 7.5.2 the proposal for a nanotechnology indicator will be discussed in more detail.

7.3 Germany's Requirements

Germany cannot disregard nanotechnology

As an internationally important location for technology, Germany cannot disregard key technologies such as nanotechnology. The industrial site of Germany itself does benefit from many raw materials. Therefore, in international competition it can only sustain its position with the help of innovations. Here, the training of high-qualified employees is an important prerequisite.

State-of-the-art technologies are a central requirement for economic progress. For this purpose, Germany needs a critical mass of know-how potential, suitably qualified employees and executives and a strong basis regarding development, manufacturing, marketing and service.

With regard to nanotechnology, the following questions arise:

- Which economic, scientific and technological basic conditions will nanotechnology meet with in Germany?
- What about the economic realization of state-of-the-art technologies in this context?

At first, the R&D intensity of nanotechnology companies in Germany is regarded as a relevant indicator.

High R&D intensity at SME

The data gathered in the present study shows an extraordinarily high proportion of R&D staff in relation to the number of employees at smaller and medium-sized companies (SME) (Fig. 7.2), which may be considered as evidence for the high R&D intensity of smaller companies in this field. Not least, this relation reflects the fact that small technology companies are generally focused on one central product, which their start-up is based on and which develops to market maturity during the seed and start phase. Experience shows that in these young technology companies this is accompanied by a near total concentration on R&D-activities. In this phase, they are particularly dependent on investors. However, especially after the end of the "new-economy-hype", the raising of venture capital in Germany turns out to be extremely difficult for start-up companies. The willingness on the part of venture-capital companies to invest in start-up companies with a relatively long "time-to-market span" has declined drastically.

Great dependence on investors

High foundation dynamics to be kept up

Consequently, there is an urgent need for alternative funding sources and instruments to keep up the start-up dynamics of the past years and the jobs created in nanotechnology.

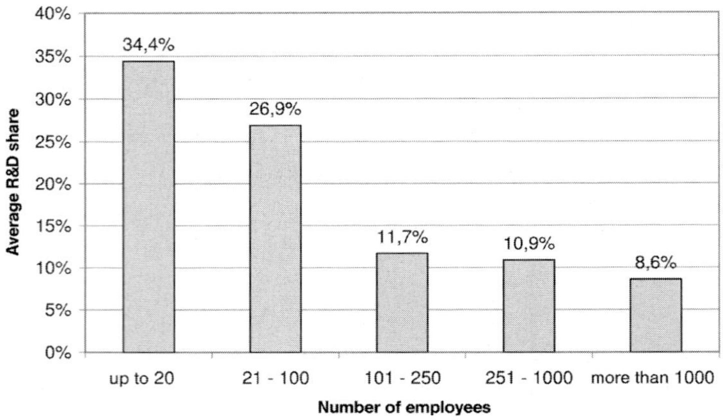

Figure 7.2 R&D share (judged by the annual turnover) of nanotechnology companies according to their size (source: paper-based company survey).

Increase in start-ups in the middle of the 1990s

Figure 7.3 gives an overview of the start-up processes in nanotechnology over the past 10 years. Here, the number of business start-ups in the year 1991 was standardized at 100 and the relative change during the following years is depicted in comparison. This development, showing dynamics comparable to other fields of high-technology (such as biotechnology), is characterized by a steep increase in busi-

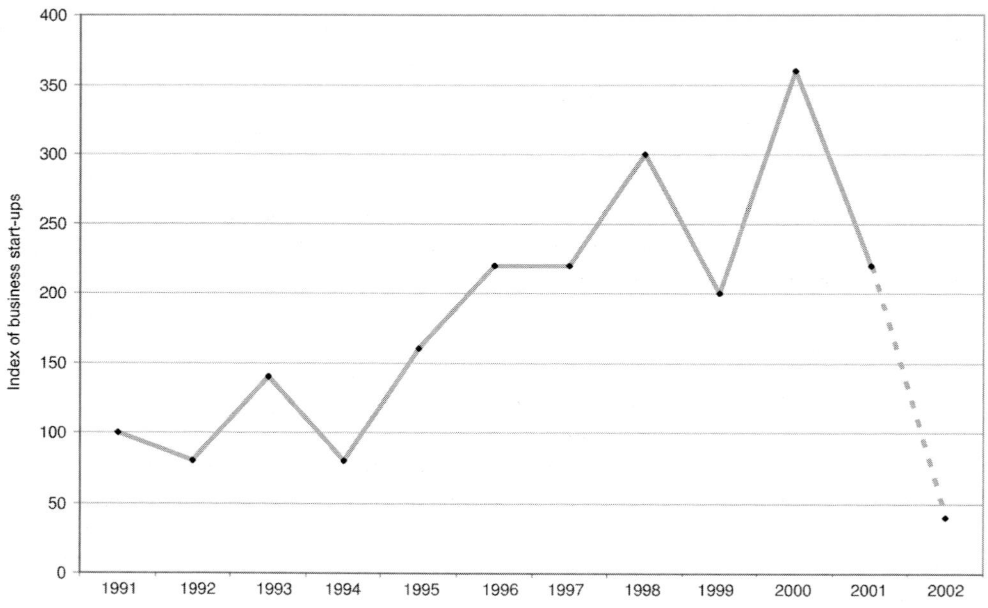

Figure 7.3 Start-up processes in Germany 1991–2002 in nanotechnology (source: paper-based company survey; Fecht et al., 2003). Start-ups in the year 1991 are standardized with 100.

ness start-ups since the middle of the 1990s, with a clear peak in the year 2000 and a significant drop in 2001. Apart from the depressed situation on the capital markets and the significantly poorer offer regarding start-up funding in comparison to the end of the 1990s, economic influences are certainly also responsible for the negative development of the start-up dynamics. The extremely low value of the business start-up index in 2002 is at least partly attributable to the incomplete recording of companies founded only recently, and is therefore only of limited informative value.

7.3.1 Technological Preconditions

Excellent basic technological conditions in Germany

Due to intensive investments in R&D and infrastructure, Germany has excellent basic technological conditions in nanotechnology. Technology indicators, such as the number of scientific publications in trade journals or the number of patents, generally reflect Germany's top position in Europe and worldwide (see e.g. TAB, 2003; Compano and Hullman, 2001).[1] The German nanotechnology scene is excellently positioned on a broad base which is reflected, inter alia, by the different topical foci of the competence centers for nanotechnology and the broadly diversified fields of operation of the companies polled within the framework of this survey. The necessity of such a broad technological positioning becomes transparent after the correlation of the functional requirements of nanotechnological products with the fields of application.

Product features relevant for nanotechnology

Table 7.1 summarizes the results in an inverse white-spot analysis of product features relevant for nanotechnology, gathered from the paper-based company survey. The white-spot analysis is a qualitative, structured and at the same time pithy method for the depiction of dependencies between markets, applications and functions or the utilization and technological platforms in the form of a matrix. Positions weakly or not at all occupied (so-called white-spots) in the matrix can be identified with the help of this method (Deschamps, 2000). If this method is used to uncover particularly *strong* connections between different facts, it is called inverse white-spot analysis. Correspondingly, the black areas represent significant causalities between functional requirements of a nanotechnological product and a specific field of application.

On the one hand, the result of the analysis shows typical functions of nanotechnological products in different fields of application, on the other hand relevant, or required functionalities and combinations of functions for a certain field of application are found. So nanotechnol-

1) For more details see Chapter 4.

Table 7.1 Inverse white-spot analysis of the most important functions and fields of application/lines of business (source: paper-based company survey).

Fields of application/ lines of business / **Functions**	ICT	Medical technology	Cosmetics	Transport	Chemistry	Environment	Energy	Mechanical engineering	Measurement technology	Building trade	Food	White/brown goods
Analytics/diagnosis		■			■				■			
Medical therapy/diagnosis		■										
Surface functionalization	■	■		■	■							
Displays	■											
Energy conversion												
Manufacturing equipment	■											
Utilization of nano/biotechnological properties		■			■							
Data processing and storage	■											
Data transmission	■											
Material separation												
Sensor technology	■	■			■				■			
Actuators												
Material metering												
Optical effects	■				■				■			
Filtering of fluids/gases												
Protection					■							
Improved material properties	■			■	■							
Structure generation	■				■							
Design / fashion / esthetics												
Protection												

ogy is applied especially in different fields where analytics and diagnosis, surface functionalization, and sensors and optical effects are important. Besides that, different functionalities of nanotechnological products are required in the fields of I&C, medicine/health and chemistry, whereas for traffic-related applications mainly surface functionalities and material properties of nanotechnology are important.

Fields with "white spots"

In the fields of cosmetics, environment, energy, mechanical engineering, building trade, food, and white/brown goods, the threshold value determined could not be exceeded in any of the application fields, so that only "white spots" are found there. This suggests that, to a large extent, the results and methods of nanotechnology are still not utilized in the fields mentioned. The results of the inverse white-spot analysis show a remarkable conformity with other sources (see e.g. Fecht at al., 2003). There, the fields of transport, energy and environment rank last regarding the influence of nanotechnology in different fields of application or lines of business. So even the TAB-study (TAB, 2003) points out that the stage of development of products, product ideas and concepts of nanotechnology – depending on the line of business – has progressed differently. The diffusion of nanotechnology in applications is still getting off the ground in many fields. The American National Science Foundation (Roco, 2002) goes on the assumption that the development of nanotechnology can be described by an S-shaped curve, and that from the beginning of the curve a period of five years is required to reach the steeply rising area. This trend can be noticed in the white-spot analyses shown in Tables 7.1 and 7.2. For a broad utilization of nanotechnology, the basis has to be created with the help of developments in the fields of material, equipment, production and analysis. When these requirements have been fulfilled, the diffusion of the developed technologies in other sectors will be possible.

Creating the basis for a broad utilization of nanotechnology

Analogously, the inverse white-spot analysis can supply information about the relevance of technology platforms of nanotechnology. In Table 7.2, the typical technology platforms as well as the dominant markets for their end products are depicted. According to the companies polled, the white squares do not represent significant technology or measurement engineering platforms for a specific field of application, whereas the black squares define important technologies, or measurement techniques for a special field of application.

The companies questioned go on the assumption that the most important fields of application of nanotechnology (I&C, chemistry and medical technology) also require most of the technology platforms and measurement techniques. With the exception of self-assembly and molecular engineering, the other technologies and measuring techniques can be classified as rather classical methods, but further developed and optimized by current nanotechnological insights. Just as in Table 7.1, the conformity of black squares with regard to the respective fields of application or lines of business can be seen in the inverse white-spot analysis (Table 7.2), too. Obviously, these are the fields in which the methods of nanotechnology are already established or being introduced. The greatest technological variety is required in the fields of medical technology, I&C and chemistry. The application of these technology platforms is a necessary basis for the manufacturing of products. A specific public funding of nanotechnology in the mentioned fields of application may accelerate the diffusion of nanotechnology in products and enhance the competitiveness of German enterprises in the short or medium term.

Table 7.2 Inverse white-spot analysis of the most important technology platforms and fields of application/lines of business (source: paper-based company survey).

Technology platform/ Measurement technology \ Fields of application/ lines of business	ICT	Medical	Cosmetics	Transport	Chemistry	Environment	Energy	Mechanical engineering	Measurement technology	Building trade	Food	White/brown goods
Separation/filtration methods												
Biological engineering		■										
Sol-gel processing				■	■							
Powder processing					■							
Particle beam lithography												
Catalysis												
Self assembly		■			■							
Nanoprint/-imprint												
Molecular engineering		■										
Metrology									■			
Microscopy	■	■			■							
Modeling and simulation	■											
Ultra-precision engineering	■	■										
Optical lithography	■	■							■			
Thin film deposition methods	■	■			■							
Others												

7.3.2
Preconditions for Industry-specific Applications

With regard to the fields of application covered by German nanotechnology companies, clear priorities within the broad field of nanotechnological products are noticeable (see Fig. 7.4).

Analogous to the intense presence of the chemical industry among the nanotechnology companies, which according to the survey represent the largest share of the companies polled with 27%, the field "chemistry" as field of application for their nanotechnological products was stated by more than 50% of the companies questioned. In second and third place, the fields of application of "information and communication technology" and "medicine/life sciences" follow, which are already serviced with nanotechnological products by almost 40% of the companies. Compared to measurement engineering, which is to be regarded as the fourth important field of application,

"Chemistry" as special field of application followed by "ICT"

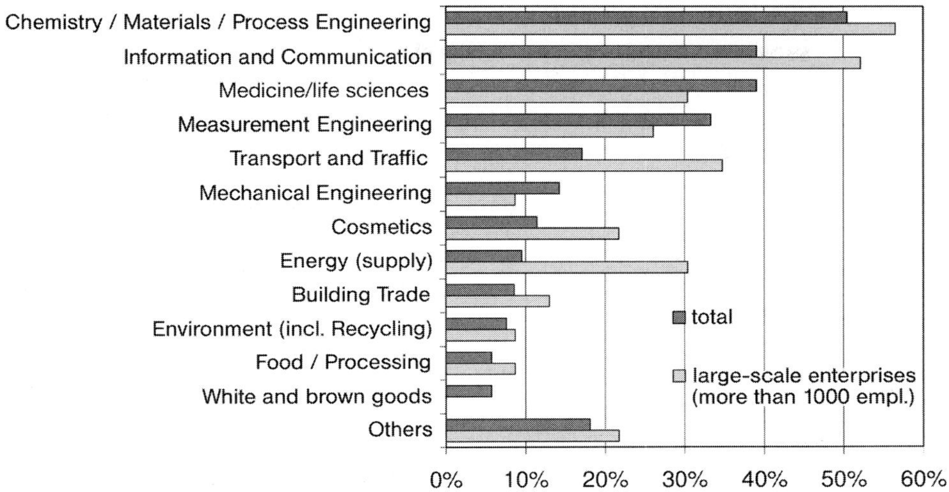

Figure 7.4 Importance of nanotechnology for individual fields of application (source: paper-based company survey).

for the majority of the companies polled the other fields are of less importance. This goes even for the field of "transport and traffic", which, however, has been defined as a Lead Market for nanotechnology owing to its overall economic importance for Germany (in this survey called "car manufacture") (see Chapter 6).

A separate analysis of large-scale enterprises with more than 1000 employees to verify the economic importance of the individual fields of application revealed some differences. A far-reaching conformity is given in the two leading fields of application, i.e. chemistry and ICT. However, as expected, a significantly higher importance is attached to the field "transport and traffic" here, in which car manufacture represents an essential sector. From the large-scale industry's point of view, the fields of application "energy (supply)" and "cosmetics" play a not insignificant role, too.

Nanotechnological know-how as an important competition factor

Moreover, the question arises as to how much technological and economical importance is attached to nanotechnology within the companies. The majority of the companies polled regard their nanotechnological know-how as an important competition factor both in their traditional markets and for the tapping of new markets (see Fig. 7.5). For only 36% of the companies, nanotechnology simply extends the existing technological competence and is not regarded as a core competence in the medium term. For less than one quarter of the companies polled, nanotechnology represents only an experimental field. These statements underline the current commercial importance of nanotechnology for the German industry. A correlation be-

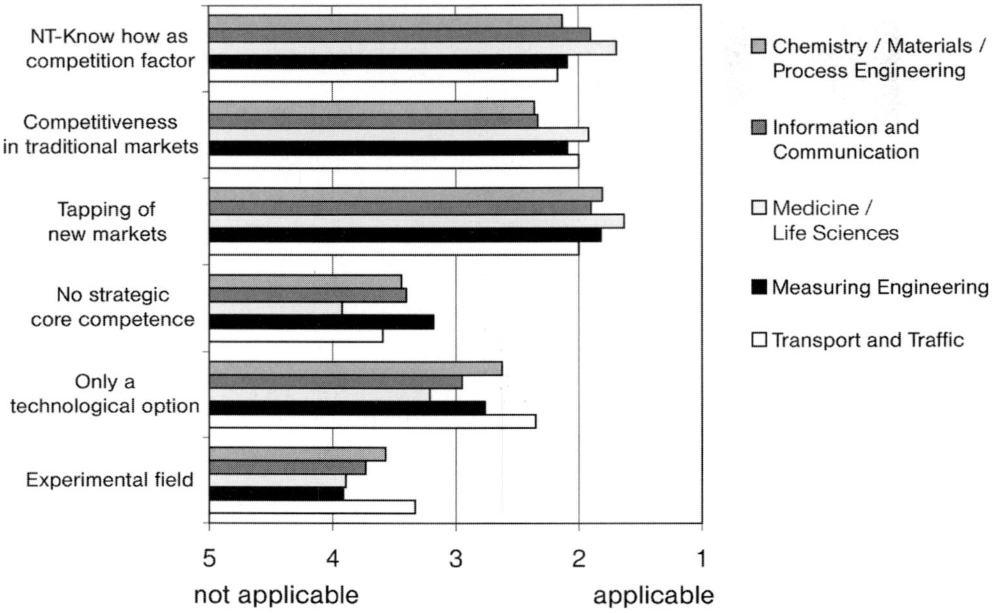

Figure 7.5 Evaluation of the importance of nanotechnology for the companies polled regarding the fields of application covered (source: paper-based company survey).

tween the statements regarding the importance of nanotechnology and the five dominant fields of application does not reveal significant differences between the fields of application (see Fig. 7.5)

A worldwide comparison of the strongest competitors of German nanotechnology companies according to their fields of application shows the strength of German enterprises especially in the application fields of chemistry and car manufacture (see Fig. 7.6). While according to German industry, Germany lies equal to the USA as the home country of the strongest competitors in these sectors, in the fields of ICT, medicine and life sciences, and measurement engineering the competitors from overseas are regarded to be stronger than Germany and the rest of the world. In the field of ICT alone, both European and Asian competitors (including Japanese) outweigh the German competition.

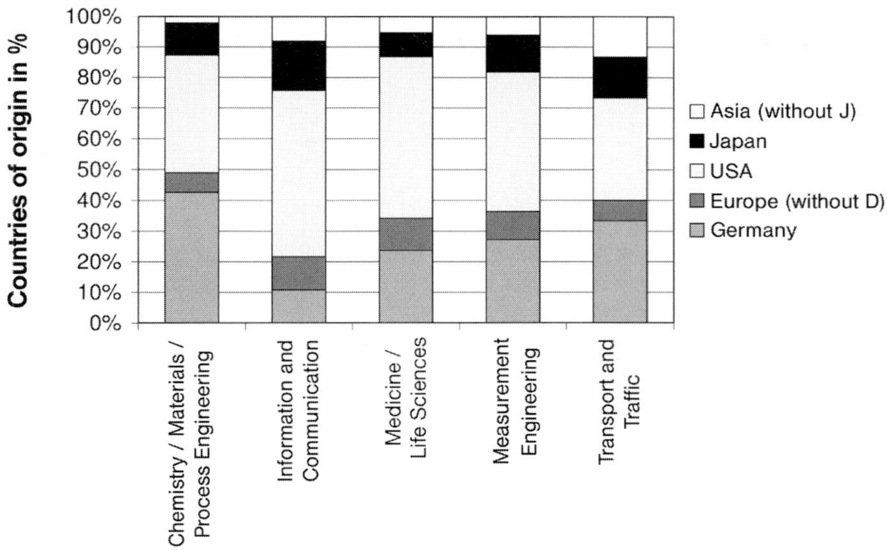

Figure 7.6 Countries of origin of the strongest competitors of German nanotechnology companies for the most important fields of application (source: paper-based company survey).

7.3.3 Financial Preconditions

As already set out, Germany has a good starting basis in nanotechnology. All the scientific technology indicators, without exception, confirm Germany's global top position (see Section 7.6). However, the technology indicators continuously reflect a period in the past, but cannot describe the current state – let alone economic development. The number of patents may form a foundation stone for future developments, however that cannot be the only decisive factor for the future economic utilization of nanotechnology in Germany. Even if the majority of the companies polled regard nanotechnology as an important competition factor, future development depends strongly on the overall economic situation.

The withholding of investments of the venture-capital businesses in start-up companies has an extremely negative effect on business start-ups in the German nanotechnological environment at the moment. The classical bank financing of business start-ups has become more and more difficult over the last years, because the German banking landscape itself is in crisis. Owing to the "new-economy hype" and the resulting negative effects for the German "venture-capital scene", new investments in risky nanotechnology start-ups become more unlikely. In addition, in Germany very few venture-capital companies have access to qualified staff who could decide reasonably on investments in such complex technologies. In this respect, some "venture-capital companies" in the USA are in a better position. In both the paper-based company survey and in all the workshops held, lack of funding sources were identified as the most important obstacle to innovation – independent from the size of the company. Funding, however, does not necessarily have to be done by "venture capital"; even in large-scale enterprises the financing of "risky nanotechnology projects" out of the "cashflow" has become more and more difficult. All in all, the funding obstacle represents a considerable problem for the future economic development of nanotechnology in Germany. In this case, not only public funding is required, but also a "reanimation" of "equity investors".

Lack of funding sources as an obstacle to innovation

7.3.4 Structure of the German "Nanotechnology Industry"

In Germany, the structure of companies active in nanotechnology shows a high number of small companies (see Fig. 7.7). This is accompanied by an age distribution of companies active in this technological field, about two thirds of which have been established since 1990 (see Fig. 7.8). The start-ups in the field of large-scale enterprises

High number of small companies

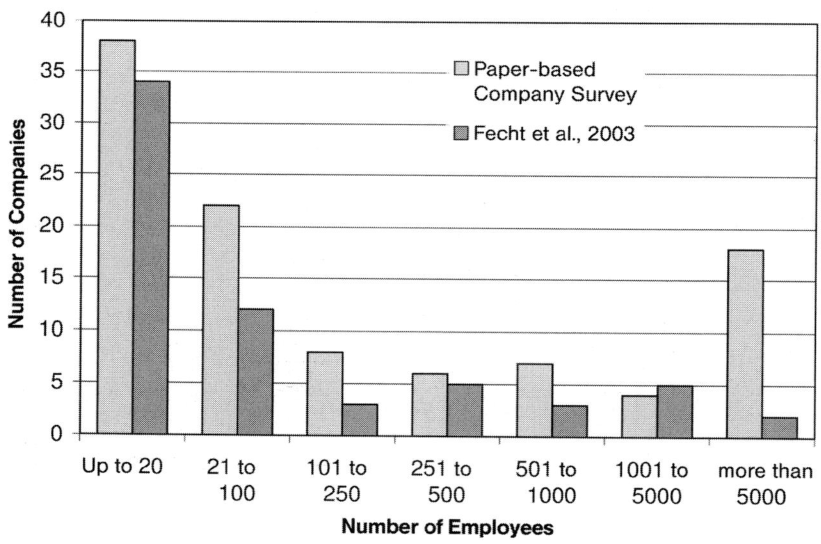

Figure 7.7 Nanotechnology companies according to size classifications (source: paper-based company survey, Fecht et al., 2003). The different sources show a tendency to conform; hence they confirm the quality of the sample.

are mainly spin-off divisions of already existing companies, such as for example the spin-off of the semiconductor division of Siemens to Infineon or the special chemistry sector of Hoechst AG to Clariant. Furthermore, Fig. 7.8 shows that the companies dealing with nanotechnology include a lot of established companies, partly founded even before 1990, such as Merck or BASF, whereas the real start-up wave of new nanotechnology companies began in the 1990s. Currently, a remarkable under-representation of medium-sized companies is noticeable in this field of technology. To a lot of smaller innovative companies, the innovation potential of nanotechnology evidently offers the opportunity to prove themselves on the market. A number of large-scale companies has also recognized the economic chances of nanotechnology and utilizes the potential of this key technology for product innovations.

Medium-sized companies under-represented

Represented large-scale enterprises very active

In Fig. 7.9, the branch assignment of the companies polled is summarized. The chemical industry and the manufacturing of measuring, control and navigation equipment could be identified as the branches of industry with the highest shares of nanotechnology companies. In addition, the latter industry shows a significant distribution frequency for SME. The NACE-coding (the National Accounts in Europe is the Eu-

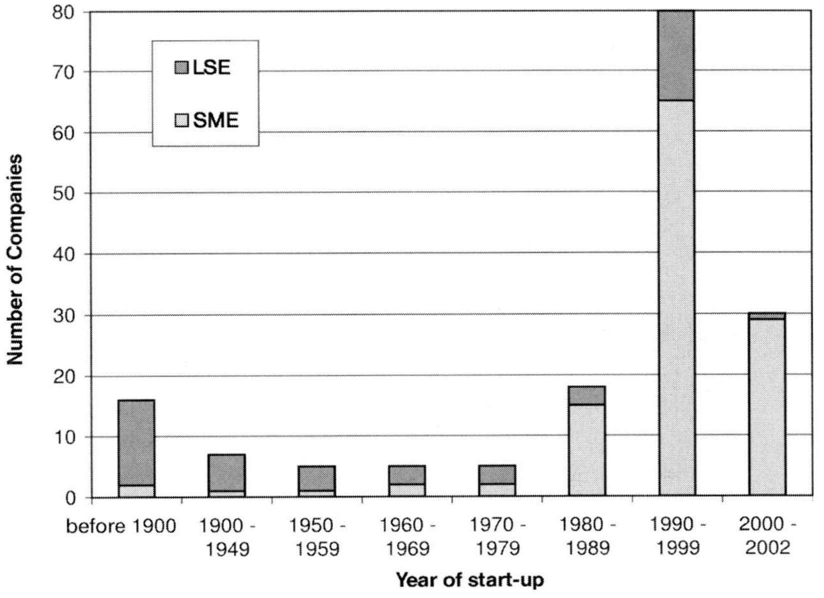

Figure 7.8 Start-up process in nanotechnology according to the size classifications of the companies (source: paper-based company survey, Fecht et al., 2003); the values for the period from 2000 to 2002 are incomplete (see Section 7.3).

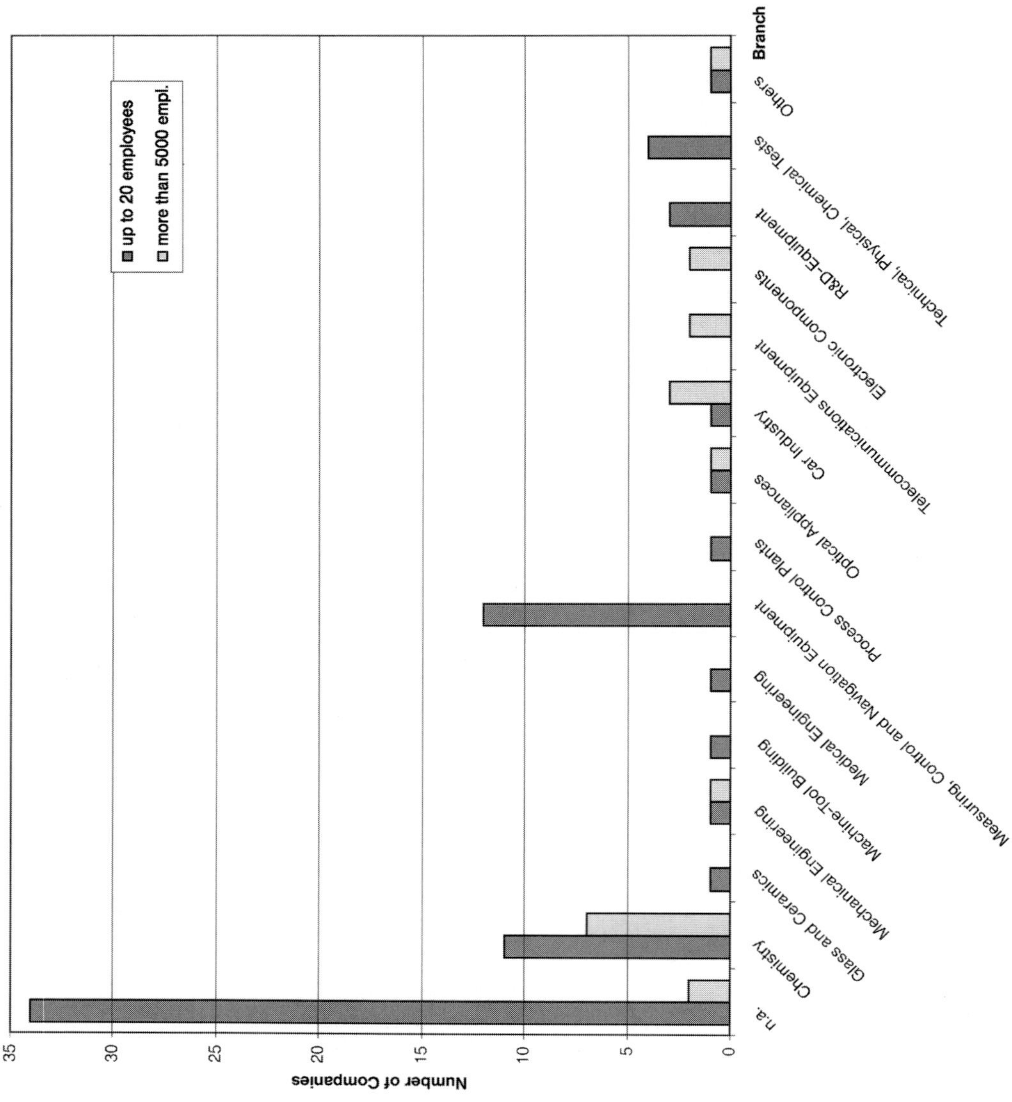

Figure 7.9 Branch assignment of the nanotechnology companies examined (source: paper-based company survey).

ropean classification of the economy into statistical sectors, respectively fields of production) used here does not differentiate clearly between classes, so it is necessary to interpret the branch assignments; it seems to show that the SME concerned are mainly active in special niches in the field of development and manufacturing of nanotechnological equipment, i.e. of components for production, analysis and testing.

7.3.5
Acceptance of Nanotechnology

Apart from the hope for significant sales potentials in nearly all industrial sectors of the economy, nanotechnology engenders far-reaching expectations, above all in the health system, the prevention of illness, care for the older people and in other fields.

Medicine: new methods

The forecast positive effects of nanotechnological developments for health and environment include;

- the development of new methods of diagnosis and therapy;
- progressive insights in biosciences;
- the understanding of biological processes;
- the development of new or the optimization of known drugs and agro-chemicals.

Nanotechnology-based diagnostic instruments may be able to detect illnesses or predispositions for illnesses earlier than ever; nanospecific metering systems may aid progress in medication. The biocompatibility of artificial implants can be improved using nanotechnological processes.

Effects on the environment may be diminished as a result of savings of material-sensitive resources; reduction of environmentally harmful by-products; decrease in energy consumption and the removal of environmentally harmful substances from the environment. This impact of nanotechnology on human health or on the relief of the environment, however, has mainly been a hypothesis up to now, and confirmation of practical application through successful R&D efforts is still only at the beginning.[2] The practical evidence of these forecasts, sometimes lacking or regarded as insufficient by the public, leads increasingly to broad discussions about the effects and consequences of the extensive use of nanotechnology – a matter not always discussed objectively.

Environment: benefits

The focus of this discussion is, inter alia, the question regarding the impact of an uncontrolled release of nanoparticles. However, the present research results in connection with this question are of limited relevance. The supposition brought into the discussion, that concerns possible negative consequences of the inhalation or cellular absorption of nanoparticles, is mainly based on analogous conclusions drawn from results of present surveys regarding the effects of ultra-

Environment: impact of uncontrolled release of nanoparticles

2) The BMBF initiated a study on this topic ("Sustainability effects through manufacturing and application of nanotechnological products"), the results of which were published in autumn 2004. The study was carried out under the management of the Institute for Ecological Economic Research, Berlin.

fine particles (e.g. asbestos). The influence of nanoparticles on reactions in the organism, for example owing to ingestion by breathing or through the skin, is still subject of research.

Nanotechnology and ethics

In the past, nanotechnology was hardly discussed in practical philosophy and ethics, and possible social impacts of its increased application were scarcely researched. Lack of information, poor communication and insufficient knowledge of the facts can lead to low social acceptance of new technologies, not least of nanotechnology, and turn into a considerable obstacle to innovation. Different social groups have demanded a moratorium on certain aspects of application of nanotechnology and research activities in this technological field (ETC, 2003) because of these existing gaps in knowledge about the consequences of nanotechnology. To avoid such "show-stoppers" and mistakes in good time, activities in the field of socio-scientific accompanying research have been initiated recently, especially in the USA, Great Britain and Germany, among others by the NSF (USA), the Royal Society (GB) as well as the BMBF and the TAB (D).

Socio-scientific accompanying research is carried out

7.4
Obstacles to Innovation and Diffusion

Against the backdrop of increasing technological dynamics in the form of shortening product cycles and technological life cycles, saturation tendencies on traditional markets and dynamization of technical developments, global competition has intensified considerably over recent years. Competitiveness and growth depend more than ever on the capability to create new processes and to market them worldwide in order to tap new markets and fulfill future requirements regarding innovation and investments. Therefore, innovation is a central determinant for competitiveness and future success, to wit not only for large-scale enterprises, but especially for SME, since if there is a U-turn in the market they are subject to more existential adjustment problems than large-scale enterprises.

Ensuring innovation-friendly atmosphere

For the international competitiveness of Germany in promising fields such as nanotechnology, it is of the utmost importance to ensure an innovation-friendly atmosphere for companies and to clear away obstacles to innovation. For this purpose, potential obstacles to innovation were recorded in the paper-based survey carried out for this study, with regard to the fields of application for nanotechnological products addressed by industry. According to the companies polled, obstacles to innovation like innovation costs, financing and public funding, conspicuously all of financial nature, rank first (see Fig. 5.41 in Chapter 5). The ranking order of the obstacles mentioned indicates that the development of new products or processes in the field of nanotechnology requires considerable investments that cannot be financed by equity capital alone.

This opinion was even confirmed by the workshops held on the chosen Lead Markets. The development of new products involves considerable risks, especially when based on new kinds of technologies not applied in the company so far. These risks raise the question of the respective cost-–benefit ratio both in case of external funding (e.g. on the capital market) and in competition with other products or company divisions regarding the allocation of internal financial resources. German companies face a generally difficult financial situation – at least partly– and a tendency towards increasingly short-term-oriented management decisions. With regard to public funding along the value-added chain, the companies lamented the fact that support was too short-term oriented.

7.4.1
Differences between SME and Large-scale Enterprises

Significant differences between the obstacles to innovation for SME and large-scale enterprises can be identified in three sectors. There is a significant difference in the sources of finance. SME regard their access conditions to the capital market as far more difficult than large-scale enterprises. Accordingly, for 38.6% of the SME the funding of their activities is an important obstacle to innovation. In contrast to this, the access to the capital market of large-scale enterprises represents an important obstacle to innovation for only 7.7% of the sample.

More difficult conditions for the access to the capital market for SME

There is a similar difference regarding access to market information. Here, too, 21.3% of the SME – a significantly higher number than for large-scale enterprises with 3.7% – regard this as an important obstacle to innovation.

Finally, the absence of competent regional cooperative partners is another obstacle to innovation the SME and the large-scale enterprises evaluate differently in the sample. 22.2% of the SME regard this obstacle to innovation as important, whereas only 7.4% of the large-scale companies consider this to be an important obstacle. Obviously, with increasing plant size, access to the capital market, the available quantity of market information and the attractiveness for regional partners also grows. For an innovation policy with the aim to strengthen SME, these differences are the basis of an approach to selectively targeted interventions.

7.4.2
Obstacles to Diffusion

The major diffusion obstacles to nanotechnology are to be found where nanotechnological concepts or primary products have not played a role so far. This applies especially to industrial sectors or lines of industry with little technological contact to the scientific basis of nanotechnology. Examples are mechanical or automotive engineering that is mainly characterized by engineering development approaches and less by scientifically-based research.

According to the car industry, reduced technical maturity is a decisive diffusion obstacle in a large number of potential fields of application, especially for the application of nanotechnology in automobiles. The comparatively early stage of development of many nanotechnological applications in cars and the still poorly established cooperative relations between provider and customer of nanotechnological products are the explanation for the fact that the market potential for the application in cars is regarded as uncertain or even below average.

Early stage of development of nanotechnology currently still a problem in some lines of industry

Even the chemical industry, which, according to the survey, is the line of business represented best among the manufacturers and users of nanotechnological products in Germany, states that the lack of information about potential markets is a barrier for the diffusion of internal nanotechnology know-how into other fields of application. Here again the position of the chemical industry, usually found at the beginning of the value-added chain, has an accordingly inhibiting effect on the diffusion of nanotechnology in end-product oriented fields of application. Further obstacles to the successful diffusion of nanotechnology into new markets and applications are, according to the chemical industry, underestimated launching periods, lack of "killer applications"[3] and the often obscure patent situation for SME, that can usually only be clarified with the help of expert knowledge and considerable resources (for more details see Chapter 4).

Diffusion barrier: lack of information about potential markets

In summary: opening up markets with the help of nanotechnology entails considerable investments that industry is not able to bear alone just like that. Limited market knowledge and as yet insufficient cooperative networks, especially in branches not yet strongly penetrated by nanotechnology, are barriers for the speed of innovation and diffusion into new fields of application. Measures to remove these barriers and the access to external capital as well as to public funding along the value-added chain are important preconditions for the improvement of the cost–benefit ratio for investments in new products and applica-

3) "Killer application" means a product able to achieve a strong market position, thus endangering the existence of already (established) competing products.

tions of nanotechnology and for the increase of investment activities in this field.

Improving measures to remove barriers

7.5
Employment Effects and Qualifications

When ascribing jobs to certain products or product groups it has to be taken into account that, in general, a certain product is not produced completely autonomously by one company in one branch of industry, but that diverse interconnections of advance transactions between companies of different branches of industry are required. Therefore, all direct and indirect employment effects, according to the following distinction, would have to be considered for a complete job analysis.

- Direct employment effects: All production activities and services connected with the development, manufacturing, marketing, sale and financing as well as with the operation of plants for the manufacturing of nanotechnological components and products.
- Indirect employment effects in the capital goods industry and service sector: To carry out development, manufacturing, marketing, sale, financing and operation, companies, authorities and other organizations involved have to invest and have to take advantage of external services. Jobs related to the production of such capital equipment and services depend only indirectly on the sales results of nanotechnological products and services.

Direct and indirect employment effects

Based on the definition of nanotechnology, which includes technical equipment and devices for analytical chemistry as well as plants for the manufacturing of nanotechnological components (e.g. coating systems), the border between direct and indirect effects becomes blurred so that, strictly speaking, even indirect effects in the form of such capital equipment are taken into account to a certain extent in the following considerations.

Furthermore, the examination of employment effects should ideally include the differentiation between gross and net effects. The gross effect describes all jobs in connection with the utilization of nanotechnology. The net effect, however, takes into account that through the application of nanotechnology other technologies are replaced so that jobs are probably lost elsewhere. Owing to the effect of nanotechnology in different stages along the value-added chain and the complexity of substitution processes related to it, consideration of the net effect within the framework of this survey was not possible.

Consideration of net effect was not possible

For the determination of employment effects, results of the paper-based company survey as well as available secondary sources, such as the Markus database,[4] company reports and websites were consulted.

Limited database

Despite the limited database, an attempt was made to assess the consolidated turnovers and the total employment figures of the German "nanotechnology industry". For these assessments, existing data was analyzed using statistical methods. To carry out the assessments, missing data of the companies included in the total sample, especially turnovers, number of employees and nanotechnological share in the turnover, was assessed on a statistical basis. It was generally assumed that a normal distribution for the given company ratios of the respective data set under review prevailed. To ensure the validity of this assumption and to increase the homogeneity within one data set, the total sample was subdivided into seven plant scale classes (according to number of employees) and statistically analyzed separately within each of these sets.

However, this method leads to a reduction of the sample size per data set and therefore operates at the expense of the statistical significance of the results. The partly very small sample sizes allow only a rough estimate, and from a statistical point of view they are to be regarded more as an indicator. On the other hand, an analysis of the whole sample would lead to far more unrealistic assessments owing to the heterogeneity of the surveyed companies and variations in the surveyed parameters (turnover and employment) of up to five orders of magnitude (decimal powers).

7.5.1
Sales and Employment Figures in the German "Nanotechnology Industry"

Basis: company survey carried out

The basis of the determination of employment effects of nanotechnology for the German national economy is the paper-based survey among the companies represented in Germany that are dealing with nanotechnology either as manufacturers or as users. Initially, the share of the companies' sales in nanotechnological products, and the share of employees directly or indirectly involved in nanotechnology, was not taken into account.

For reasons of acceptance, the interviewees of the company survey were only questioned about a categorization of the companies with regard to sales and employment according to predetermined size classifications. In order to enable more precise statements about the number of employees, absolute employment figures from other available sources were used, inter alia the Markus database, from published an-

4) The Markus-database is a commercial database of the organisation Creditreform containing about 800 000 entries on German companies (inter alia, object of business, turnover, number of employees

nual reports and occasionally from company websites. These figures served to validate company data given in the questionnaire and were included in the model calculation for the assessment of the total employment figures.

Additional source: Markus-database

In addition, a current commercial study on nanotechnology (Fecht et al., 2003), independent from this survey, was used to increase the parent population for the assessment and to receive a representative statement regarding the employment effect of nanotechnology in Germany. In this report, a total of 99 German "nanotechnology companies" was represented, for the most part with verified statements regarding sales and employment figures. A comparison of both data stocks showed that for the following consideration an extended sample of altogether 167 companies (total sample) could be used. Table 7.3 summarizes the company samples serving as a basis for the following assessment.

Additional source: commercial study on nanotechnology

Table 7.3 Sample for the assessment of turnovers and employees.

	Companies of source data acquisition (paper-based company survey)	**Additional companies from secondary sources (Fecht et al., 2003)**	**Total sample of companies**
Number of companies	103	64	167
with sales figures	73 (71%)	48 (75%)	121 (72%)
with employment figures	82 (79%)	64 (100%)	146 (87%)

The database for sales and employment figures of 72%, and 87%, and allows a confident extension to the total sample of 167 companies. For this purpose, the companies were classified by size, since for large-scale enterprises with more than 5000 employees, the sales and employment figures of almost all of the companies included in the sample were available, whereas sales figures of SME with up to 20 employees were available only in just over 60% of the cases. For the projection, the statistical mean per size classification was used. A summary of the present data and of the projection to the total sample is given in Table 7.4.

Sample used

The 167 companies of the sample that are active in nanotechnology made an estimated total turnover (worldwide) of almost 360 billion EUR with a total number of 1.26 million employees, in which the sales share through nanotechnology is not yet determined.

Table 7.4 Assessment of sales and employees of 167 "nanotechnology companies" (concerning the segmentation of the companies in F1, F2 and F2, see Section 7.2.2) in 2002 (total sample).

Size classification	**1**	**2**	**3**	**4**	**5**	**6**	**7**	**Total**
Number of employees	up to 20	21 to 100	101 to 250	251 to 500	501 to 1 000	1 001 to 5 000	more than 5 000	
Number of companies	72	34	11	11	10	9	20	167
Assessment total sales (in million EUR)	79	197	258	950	1 109	5 038	353 072	360 703
Assessment total number of employees	650	1 382	1 761	3 944	7 161	23 463	1 223 837	1 262 198

Although the projected sales and employment figures depicted here can be regarded as absolutely representative owing to the good coverage of the sample, they should be interpreted with care. As already established, these figures refer to the respective company as a whole, without considering the sales share attributable to nanotechnology. With this, Table 7.4 completely contains the total sales and employment figures of the large-scale enterprises included in the sample, such as DaimlerChrysler, Bayer or BASF. For a further differentiation of nanotechnological relevance, the estimated sales share of nanotechnology in the total sales of the respective company was asked about in the company survey. 73% of the companies polled in the source data acquisition ($n = 103$) answered this question. The results, subdivided according to the company size, are shown in Fig. 7.10. For example, 15% of the companies with more than 1000 employees say that the share of nanotechnology in their total sales is below 1%.

There is a significant correlation between company size and the share of nanotechnology in the company. This is no surprise, since small companies have to concentrate their resources more intensively, and therefore use a more strictly-defined core competence than large-scale companies. In particular, young nanotechnology companies stand out due to a pronounced competence in a special branch of nanotechnology, and they use this know-how as a success and competition factor to position themselves successfully on the market.

It is noticeable that among smaller companies the share of nanotechnology varies more greatly than it is the case with large-scale enterprises. This may partly have to do with the fact that SME dominate the sample, thus show a higher variance. The clear trend that with an increasing company size, the companies deal either a) predominantly

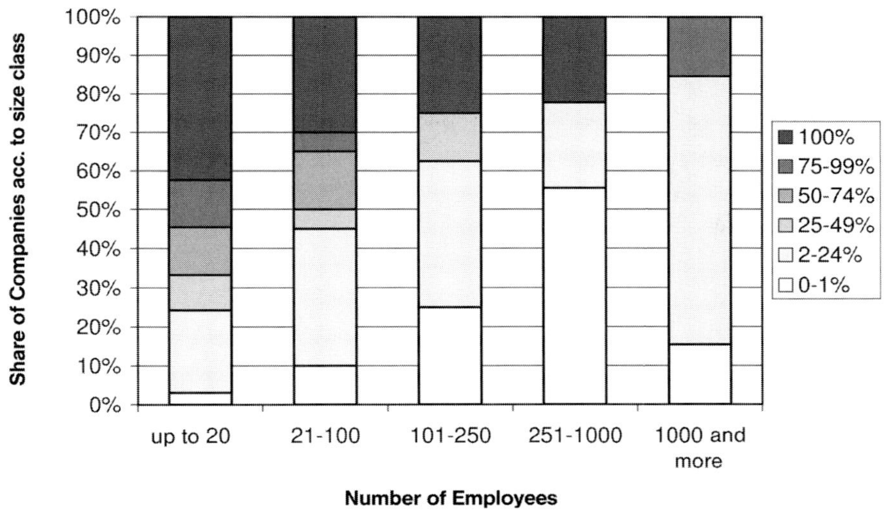

Figure 7.10 Sales share of nanotechnology in the total sales according to company size (source: paper-based company survey).

or b) only in passing, with nanotechnology, explains that there are obviously further substantial reasons.

One reason is the heterogeneity of the markets for nanotechnological products regarding their maturity and sales volume. Nanotechnology serves both established mass markets and potential growth markets or selected niche markets. For the industrial enterprises that usually act worldwide, however, only large established markets or developing markets with promising growth potentials are worth considering. In the first case, a high sales share of the company is attributable to nanotechnology, whereas in the latter case, nanotechnology sales are still of little importance for the companies – a fact that explains the discrepancy between small and large-scale enterprises in the evaluation of nanotechnology.

Against this backdrop, Fig. 7.11 reflects the existing and anticipated potentials of nanotechnology regarding the industrial structure of Germany. Here, it becomes clear that among the major nanotechnology companies, the chemical industry dominates, followed by optics and the car and electronics industries.

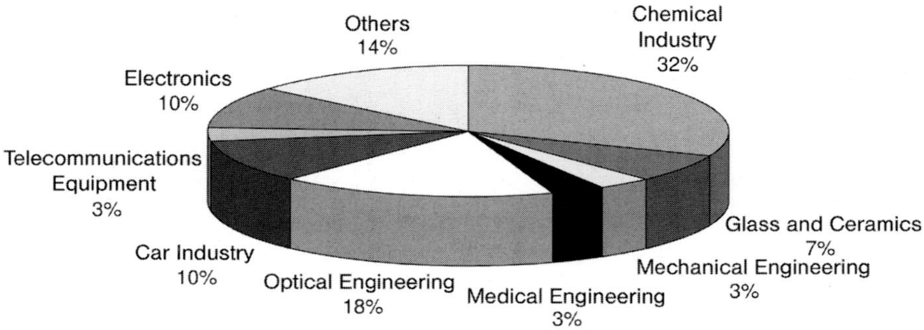

Figure 7.11 Industry-specific distribution of nanotechnology companies with more than 1000 employees ($n = 29$). (Source: paper-based company survey).

7.5.2
Assessment of Employment Effects of Nanotechnology in Germany

Preliminary assessment of employment effects

It is only possible to make a preliminary assessment of the employment effects in Germany related to nanotechnology, within the framework of this survey, because the available database is incomplete and there are problems associated with gathering reliable figures. It was not possible to determine the actual employment figures exactly because of the cross-disciplinary character of nanotechnology; the fact that this technological field is not incorporated into the statistically recorded systemizing schemes; the inclusion of the overall economic production statistics; and the application of complex input–output models. Nevertheless, the assessment carried out here is based on approx. 37%[5)] of the nanotechnology companies operating in Germany. Therefore, the results can be regarded as absolutely reliable guideline values.

Notes on the quality of the database

Prior to explaining the results of the following assessment, here are some notes on the quality of the database, the assumptions and the chosen methodology in order to ensure a comprehensive transparency of the basic prerequisites

- To determine the employment figures, the sales share of nanotechnology was used and transferred in a ratio of 1:1 to the proportion of employees related to nanotechnology; this tends to result in a lower number of employees than the actual one since, according to the given information, nanotechnological products are still in the stage of development in many companies and have not achieved any turnover yet.

5) The number of companies in Germany known, or at least presumed, to operate currently in the field of nanotechnology is approx. 450.

- Information about the sales share of nanotechnology were only available for 44% of the companies of the extended sample ($n = 167$); the validity of this information could not be verified.
- It was only possible to differentiate between employees in Germany and foreign employees of the companies polled for the 20 large-scale enterprises with more than 5000 employees included in the sample.
- No qualitative differentiation was made according to the kind of employment, for example according to highly-qualified researchers and development engineers in comparison to simple jobs in the production sector.

For this assessment, all available company data and relevant parameters (sales, employees and nanotechnology share) of the companies were taken into consideration in order to ensure the highest possible reliability of the assessment based on the available data (see Table 7.5).

Based on this assessment, a worldwide total nanotechnology turnover of approx. 30 billion EUR and a number of employees of 69 000 (direct or indirect employment calculated with the help of the sales figures) arises for all 167 companies surveyed, in which large-scale businesses with more than 5 000 employees have by far the greatest influence on the totals sales (approx. 95% respectively 89%) and the total employment. One reason for this is the fact that, according to their own information, some large-scale enterprises have a high proportion of nanotechnology and consequently, their number of employees is brought into the assessment to this extent. This goes mainly for large-scale enterprises of the chemical or optical industry that already achieve sales with established product lines that run into billions, such as nanostructured materials (e.g. carbon black or polymer dispersions, see Chapter 6.2) or nanooptical components (lithograph-

Table 7.5 Assessment of sales and employees related to nanotechnology, in Germany in 2002 for the total sample.

Size classification	1	2	3	4	5	6	7	Total
Number of employees	up to 20	21 to 100	101 to 250	251 to 500	501 to 1 000	1 001 to 5 000	more than 5 000	
Number of companies	72	34	11	11	10	9	20	167
Estimated nanotechnology sales (in million EUR)	46	93	92	263	246	1 211	28 353	30 304
Estimated nanotechnology employees	302	511	407	746	1 209	4 313	61 603	69 091

ic optics). In order to take into account that mainly in larger enterprises a significant part of the performance occurs abroad, without having an effect on the domestic employment in Germany, the domestic employment rate for the determination of the domestic employment effect induced by nanotechnology ought to be considered to the greatest possible extent. Therefore, for the 20 companies with more than 5 000 employees in the sample the share of employees in Germany was determined additionally and applied analogously to the potential employees in nanotechnology.

The results of the assessment shown in Table 7.5 permit the conclusion that nanotechnology is to be found in the particularly innovative value-added activities of the companies, and that these activities tend to be more established in Germany owing to the high level of qualification of the local employees and the proximity to the central R&D locations of the German enterprises.

In the following, the currently estimated employment figures are projected in three steps to 450 companies operating in the field of nanotechnology. In the first step (see Fig. 7.12), the employment figures of the companies with up to 5000 employees were extrapolated to altogether 420 companies. This results in a number of employees of 21 400.

Additional extrapolation

In the case of companies with more than 5000 employees, only a conservative extrapolation from 20 to 30 companies was carried out (Fig. 7.13). This resulted in a number of employees of 92 400. For the

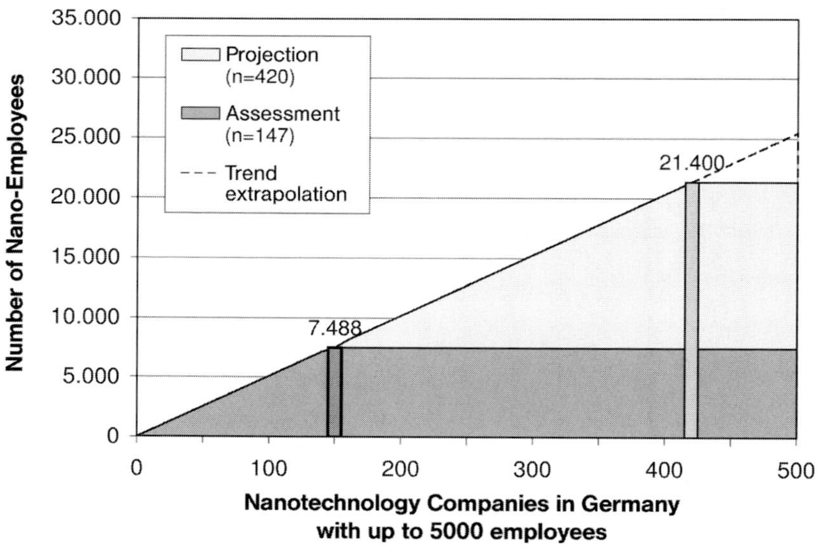

Figure 7.12 Extrapolation 1 with regard to companies with up to 5000 employees.

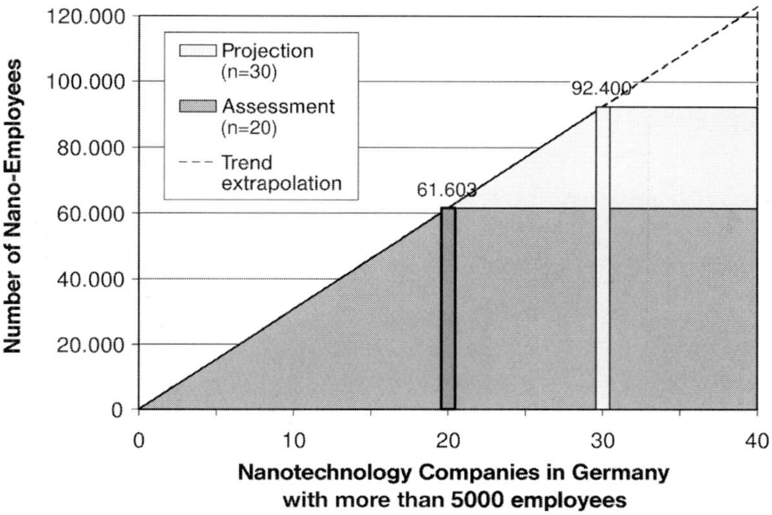

Figure 7.13 Extrapolation 2 with regard to companies with more than 5000 employees.

authors of this study, this procedure was plausible, due to the comparison of the companies having answered the questionnaire with the list of addresses of companies with more than 5000 employees.

In Fig. 7.14, extrapolation 1 and 2 are summarized. With regard to the 450 companies in Germany that are known to operate in the field of nanotechnology at the moment or presumed to do so, it can be established that according to this calculation based on sales figures 114 000 jobs are to be estimated.

Optimistic estimate: currently 114 000 jobs

For the validation of the assessment of employment effects of nanotechnology, company information of the paper-based survey carried out within the framework of this study was utilized in a further step, in which the company was questioned about the employees depending directly or indirectly on nanotechnology (see Q.14 in the questionnaire and Fig. 7.15). According to this, the 92 companies answering this question employed at least between 4000 and 6500 people in this field. The gap of the calculated minimum number of employees results from the fact that size ranges were asked instead of absolute figures.

Furthermore, the values are to be regarded as the lower limit of employment induced by nanotechnology, since in this calculation companies with more than 250 employees depending on nanotechnology can be taken into account only up to this value.

After linear interpolation, assessment of the minimum employment effect of nanotechnology for the approximately 450 "nanotech-

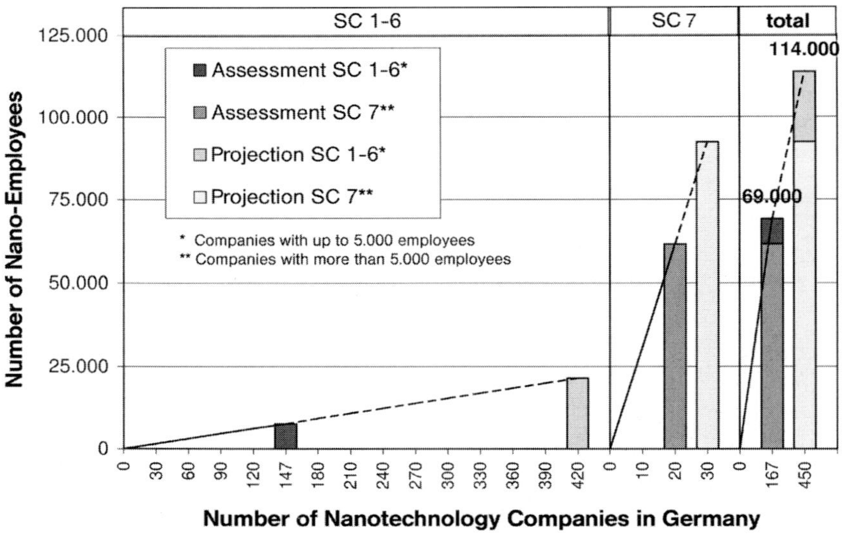

Figure 7.14 Summary of extrapolation 1 and 2 as well as the number of employees in 450 "nanotechnology companies" in Germany.

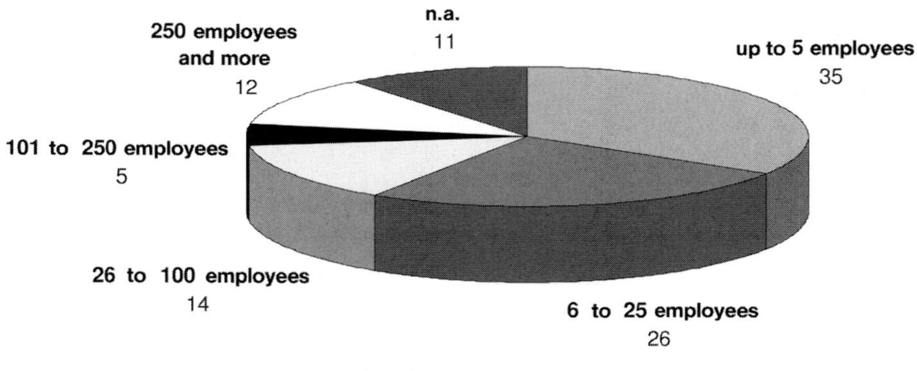

Figure 7.15 Number of employees in the company directly or indirectly depending on nanotechnology (total 103 companies, source: paper-based company survey).

Conservative estimate: currently 20 000 to 32 000 jobs

nology companies" in Germany on the basis of this data amounts to a number of at least 20 000 to 32 000 jobs in Germany. The determined total numbers of employees that arise depending on the respective arithmetical approach are compared in a graph in Fig. 7.16.

Depending on database and methodology used, extremely different estimates regarding the total employment in connection with nanotechnology are reached. As already mentioned, the range of the lower estimates of 20 000 to 32 000 has to be interpreted as a lower limit.

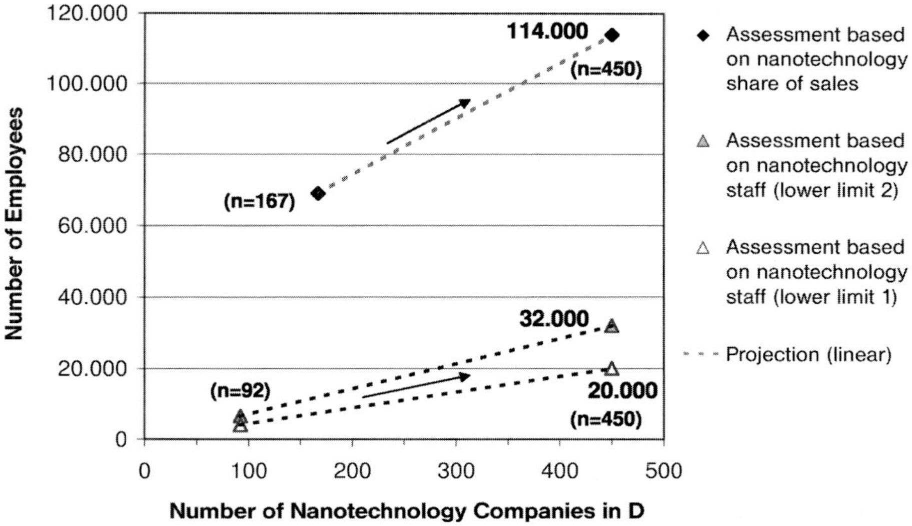

Figure 7.16 Comparison of the projected employment effect of nanotechnology for Germany in the year 2002 according to different assessments (own calculations).

Consequently, the 114 000 jobs estimated on the basis of the sales figures can be seen as an upper limit. With regard to the employment effect, it has to be taken into consideration that this stated number of jobs is mainly established in the large-scale industry. These figures do not contain jobs in non-industrial fields as they are to be found in scientific institutions and in the university and college system.

Lower limit: 20 000 to 32 000 jobs; Upper limit: 114 000 jobs

7.5.3 Employment Trend

Apart from the assessment of the absolute employment effects of nanotechnology, the dynamics of the employment trend is of great interest in order to be able to make statements about the development of jobs depending on nanotechnology. In this context, both the development in the past and the potential future prospects are interesting.

Just as in the case of the statistical examination of employment, even in this case only assessments on the basis of the available database are possible. For this purpose, established numbers of nearly 60 companies with regard to the number of employees in the respective company over the years 2000 to 2002 were available at the fixed date of the respective financial year. For the year 2003, available data regarding the expected number of employees of about 30 enterprises was utilized to depict an approximate trend for the past year. Just as

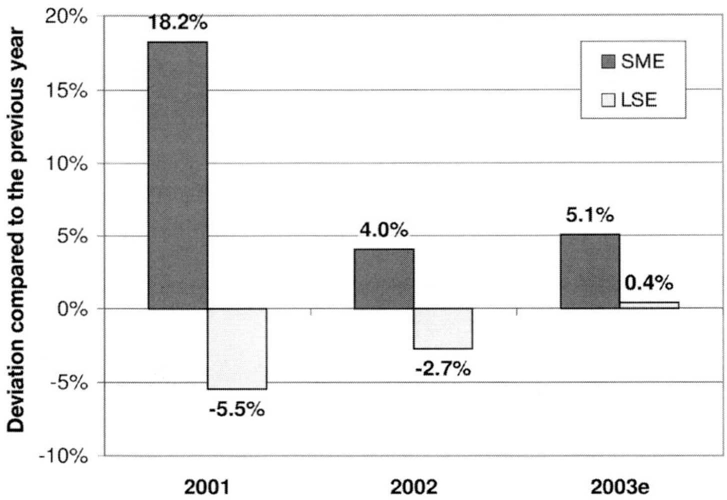

Figure 7.17 Development of the total number of employees of nanotechnology companies, separated according to SME and large-scale enterprises from 2000 to 2003[6] in the total sample.

for previous analyses, these figures were taken from the Markus database and the companies' published annual reports.

In Fig. 7.17, a graph shows the development over the past three years separated according to SME (less than 250 employees) and large-scale enterprises (more than 250 employees). The medium range hardly included in the sample (see Fig. 5.8 in Chapter 5) is not taken into consideration at this point. While all enterprises with less than 250 employees were able to record an increase in employment, the number of employees in the companies with more than 250 employees decreased in 2001 and 2002, each compared with the previous year. The positive development in the year 2003, especially of the large-scale enterprises (LSE) has to be examined with great care, since this data was gathered from a significantly smaller sample.

Concerning the SME, a significant drop occurred in the employment rise from about 18% in the year 2001 to 4 or 5% in the years 2002 and 2003. This is mainly attributable to the "burst of the New Economy bubble" in the year 2000 accompanied by the decline in venture capital for young high-tech companies. The slight delay in time between the collapse on the capital markets and the development of the employment figures in SME can be explained by the good capital stock of the companies persisting at that moment after a successful financing round.

Most companies polled expect employment growth

6) The 2003 figures are based on the numbers of employees expected by the companies.

For the assessment of the medium-term employment trend in industry influenced directly or indirectly by nanotechnology, the companies were questioned about their expectations (see Q.20 in the questionnaire). Only 21.5% of the SME and 13.5% of the large-scale enterprises do not expect an increase in personnel in this field until 2006. However, altogether 81.4% of the companies polled reckon on a more or less high increase in staff through nanotechnology in their own company.

Hence, nanotechnology can certainly be granted a high potential concerning the creation of new jobs. Almost half of the SME reckon on an increase in staff of more than 20%, one third expects more than 50% and almost 20% of the SME expect even a growth of more than 100% in this field until 2006 (see also Fig. 5.26 in Chapter 5). The forecast of the large-scale companies is not so optimistic. After all, more than half of them go on the assumption that there will be a growth of more than 10%. Here again, there are clear growth potentials in the field of nanotechnology. In the case of large-scale enterprises, however, it can be assumed that the number of employees in nanotechnology is more likely to increase owing to in-house reorganizations or conversions than through the creation of additional jobs.

Based on the additions to workforce in the field of nanotechnology expected by the companies polled, an assessment of the absolute increase in employment was made. For this purpose, the determination of the current employment was based on the same assumptions as the conservative estimate in Section 7.5.2 (ranging from 20 000 to 32 000 and 114 000). According to that, up to 2006 an expected absolute increase in employment of at least 2000 to 10 000 additional jobs will arise for the approx. 450 nanotechnology companies currently existing in Germany alone.

Increase in new jobs of 10 000 to 15 000 in the near future (2006)

Based on the same assumptions as the assessment of Section 7.5.2, a growth in newly created jobs through nanotechnology in the order of magnitude of at least 10 000 to 15 000 is to be expected.

7.5.4 Qualifications

The aspect of employment is closely connected with the issue of qualification. Usually, the increasing employment in the field of nanotechnology is accompanied by the need for qualified skilled personnel. Therefore, today many branches of industry have a demand for natural scientists with sound expert knowledge in the field of nanotechnology.[7)]

7) The BMBF is currently supporting a project on the subject "Determining of trend qualifications in the field of nanochemistry/materials and nanoanalytics". The Institut für Strukturpolitik und Wirtschaftsförderung (ISW), Halle, Germany; is entrusted with the preparation of the study. Results of the project were published at the end of the year 2004.

Nanotechnology is an interdisciplinary scientific field in which insights of physics, chemistry, engineering sciences, material sciences, biology and other disciplines meet. Therefore, innovations in the field of nanotechnology require more than expert knowledge in only one discipline. A reason for this is the blurring of the traditional borders between the natural sciences on a nuclear level. Consequently, companies operating in the field of nanotechnology have to acquire personnel who know the respective aspects of different disciplines and know how to apply their methods.

Up to now, graduates of different subject areas have been brought in for the different fields of nanotechnology. In the field of "ultraprecise surface treatment", these are mainly physicists and mechanical engineers, in "nanomaterials research" mainly chemists. Work on "biological detection systems" requires biologists, for the development of these systems physicists are needed. In the field of "functional coating systems" physicists and chemists usually work together, here however, even biologists are more and more needed.

New courses

Since the academic year 2000/2001, German universities and colleges have offered the first courses in the field of nanotechnology. Pioneers are the University of Würzburg with the course in "Nanostructure Engineering" and the University of the Saarland with the course in "Micro and Nanostructures". Further examples are the University of Kassel, since 2003 offering the interdisciplinary course of "Nanostructure Science", and the College of Applied Sciences in Munich with a course in "Micro and Nano Engineering" as master's degree courses, the latter of which is not intended for first-year students. According to the cross-sectional character of nanotechnology, these courses of study are interdisciplinary-oriented and include several faculties such as biology, physics, chemistry, material sciences or electrical engineering. In some of the courses special significance is attached to the practical orientation and the applicability of the education, and they provide for a final qualification with a degree in engineering. In addition to that, the subject nanotechnology already plays a central role in a large number of conventional courses of study like mechanical and electrical engineering, electronics, physics, chemistry or biology.

Main fields of research at universities and other centers

Besides, nearly all universities and more and more colleges of applied sciences with technical/natural-scientific courses now conduct research in the field of nanotechnology (see. Cebulla, Malanowski and Zweck, 2006). Examples are, inter alia, the Universities of Karlsruhe, Aachen, Munich, Münster, the Saarland, Kaiserslautern, Berlin, Kassel, Würzburg and Marburg. Internal decentralized networks, such as the Center for Nanoscience (Munich), the Center for Interdisciplinary Nanostructure Science and Technology (Kassel) and the Center for Functional Nanostructures (Karlsruhe) carry out independent re-

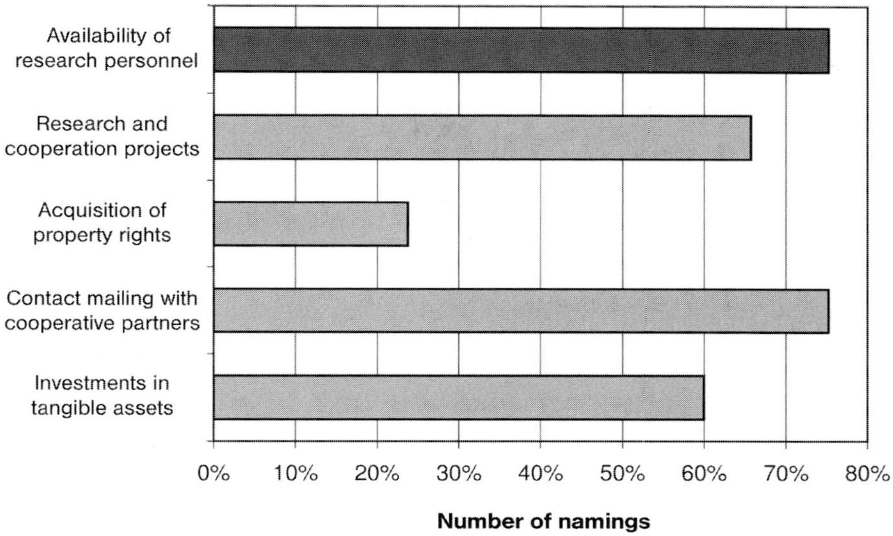

Figure 7.18 Steps taken by companies to realize nanotechnological projects for the future (source: paper-based company survey).

search work. So-called incubators and foundation centers for nanotechnology are built in the environment of universities, such as the "Center for Nanotechnology" in Münster, which support spin-offs especially in this field.

In the company survey, even different aspects regarding the issue of qualification were questioned that give information about the current need for qualified skilled personnel in industry. Regarding the question of which steps have already been taken to develop new applications or to substitute existing products with the help of nanotechnology, the availability of research personnel – apart from contact making with cooperative partners – ranks among the top priorities (see Fig. 7.18).

Need for qualified skilled personnel in the industry

At the same time, the availability of suitably skilled personnel for future projects in the field of nanotechnology was not regarded as one of the main obstacles to achieve the anticipated application and market potential. Although the lack of qualified skilled personnel was mentioned in fourth place, it was only evaluated as of average importance.

On a closer and more discriminating examination of the evaluation of this obstacle to innovation classified by the fields of application aspired to, and by the industrial origin of the interviewees, there are some idiosyncrasies (see Figs. 7.19 and 7.20).

Particularly in the field of "chemistry", the lack of suitable and skilled personnel is of below average importance. On the logical as-

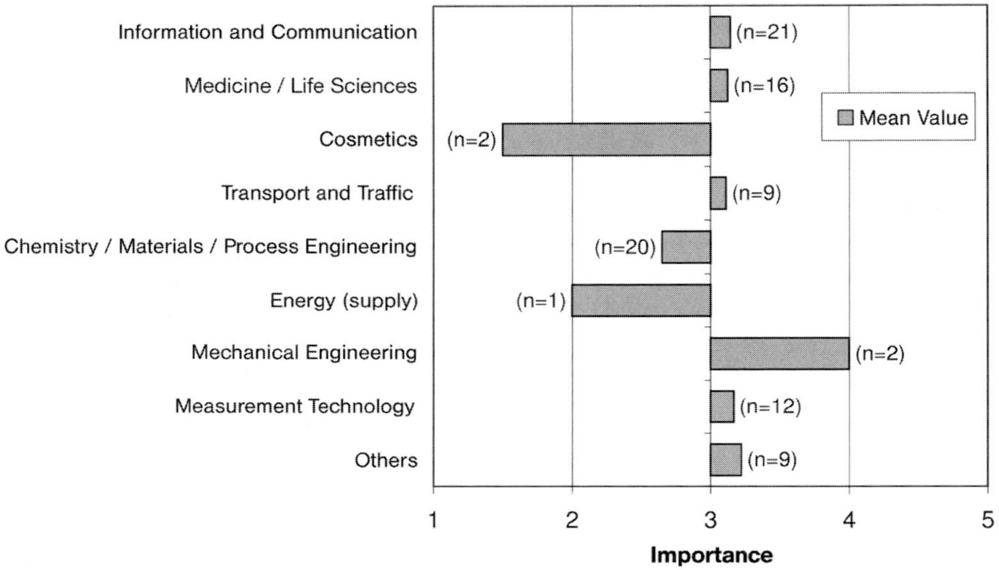

Figure 7.19 Deviation from the average in the assessment of the obstacle to innovation "skilled personnel" according to fields of application (1 = unimportant, 5 = important) (source: paper-based company survey).

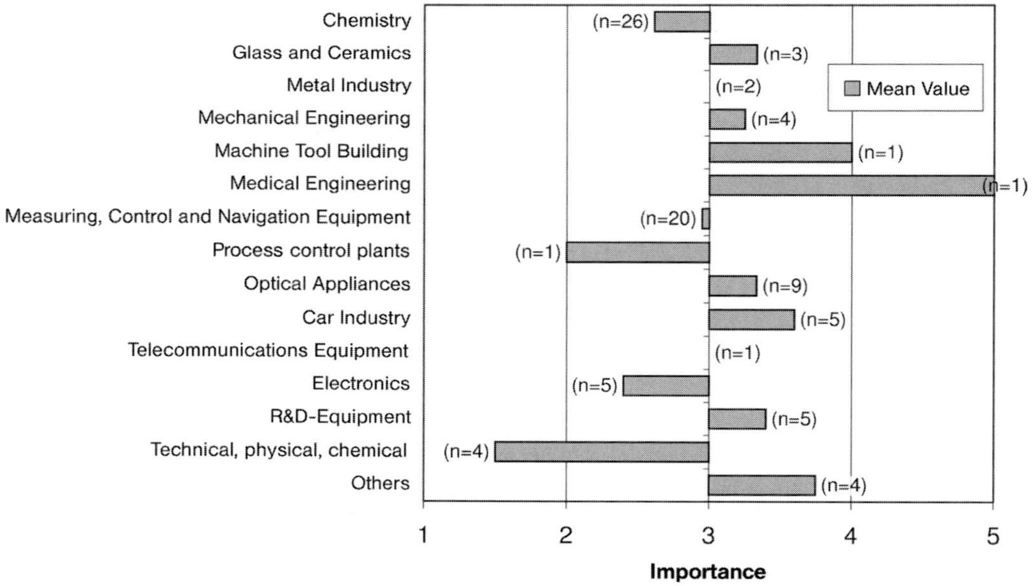

Figure 7.20 Deviation from the average in the assessment of the obstacle to innovation "skilled personnel" according to their industrial sector (1 = unimportant, 5 = important) (source: paper-based company survey).

sumption that enterprises wish to apply future nanotechnological products primarily in fields of application already known, these observations can be explained by the fact that in nanotechnology, central importance is attached especially to materials. Therefore, it is to be assumed that employees in such companies already have the appropriate knowledge with regard to nanotechnological aspects or that companies at least have access to such personnel.

Field of chemistry: does not lack suitable, skilled personnel

However, in the fields of mechanical engineering and measurement technology, this cannot necessarily be taken for granted, as the existing staff are mainly engineers whose practice-oriented education may not have included training in the basic disciplines of physics, chemistry or biology. Accordingly, the lack of suitable, skilled personnel was regarded as being more significant.

However, a representative statement for all fields of application cannot be made due to the limited scope of data and the subjective standard of assessment. The same goes for the assessment according to the industrial origin of the companies polled. An additional evaluation of the situation regarding the qualification in the Lead Markets was achieved within the context of the expert workshops held.

In the field of optics, the industrial enterprises represented did not see any need for new job outlines, such as a graduate engineer in nanotechnology. Only special additional qualifications based on a thorough basic education were regarded as appropriate. There is a continuous transition from nanooptics to microsystem engineering, for which a lot of qualified courses of training and study have recently been created so that an additional nano-course is not considered to be necessary. Furthermore, nanooptics uses an important interface between material sciences and precision treatment, so that qualifications related to them are also suitable for the field of optics.

Field of optics: Additional qualifications desired

In car manufacture, the recruitment of qualified staff is not regarded as a problem either. A bottleneck in qualified skilled personnel is found in medium-sized companies. A thorough basic education in the fields of physics, chemistry and biology is regarded as a necessary precondition for a job at the interface between nanotechnology and car engineering. Forming teams may be an approach to ensure the required interdisciplinarity and to lift communication barriers between the disciplines. Since this approach is often impracticable in SME, owing to limited personnel resources, possibilities for further education as well as nanotechnological specializations integrated into training and education should be offered.

Car manufacture: Further education and nanotechnological specialization advantageous

In summary, it can be said that the range of courses of study and training in the field of nanotechnology corresponds to a large extent to the requirements of industry. The increasing demand for interdisciplinarity of education and training is met by the range of integrated specialization areas offered in the field of nanotechnology. However,

Currently offered range of courses of study and training largely correspond to the requirements of industry

this should not happen at the expense of the basic education and training in classical sciences regarded as being of vital importance by all branches of industry. New initiatives for the support of basic education and training in connection with nanotechnology have recently been promoted by the BMBF (BMBF, 2004).

7.6
Germany's Position: an International Comparison

7.6.1
Evaluation of German Companies

To exploit the economic potential of nanotechnology and the formation of a "nanotechnology industry" that triggers off clear employment effects, one of the decisive factors is a top position in the global competition both technologically and with regard to the market. To evaluate Germany's position in the field of nanotechnology, a large number of existing surveys and studies were used (inter alia 3i, 2002; Compano and Hullmann, 2002; Fecht et al., 2003; TAB 2003). They are partly based on measurable indicators, such as the number of patents or scientific publications (Compano and Hullmann, 2002), and partly on assessments of a larger number of experts or companies questioned about the scientific or economic ranking order of the leading industrial countries, including Germany, in the field of nanotechnology (e.g. Fecht et al., 2003). An overview of the results of these surveys is given in Table 7.6.

Germany takes leading position in Europe

A comparison of different technology indicators and subjective rankings over different periods of observation shows clearly that Germany takes the leading position in Europe.[8] Only in the case of growth rates of publications and patents, Germany ranks 13th and 4th (see TAB, 2003), a fact that emphasizes the high standard of nanotechnological activities already achieved in Germany. For example, countries such as South Korea, starting from a significantly worse position, find it relatively easy to increase their growth rate. On the other hand, this shows that even smaller and economically less important countries regard nanotechnology as a promising field for themselves. With the growth rates left aside, Germany takes third place worldwide on average for the rest of the indicators.

This is also supported by the assessment of the companies questioned for this survey about the international position of nanotechnology. Asked for their assessment with regard to research and commercial realization of nanotechnology in their own fields of operation,

8) As set out in detail in Chapter 4, Germany ranks first in the world with regard to patents, almost equal to the USA.

Table 7.6 Germany's position with regard to nanotechnology in the international comparison.

Technology indicator	Period or point in time	Position (rank) of D in the international comparison	Leading country worldwide	Leading country in Europe	Source
Developments in nanotechnology (survey) medicine/pharmacy	2002	4	USA (West)	UK	3i, 2002
Developments in nanotechnology (survey) materials	2002	4	USA (West)	D	3i, 2002
Developments in nanotechnology (survey) chemicals	2002	1	D	D	3i, 2002
Developments in nanotechnology (survey) electronics	2002	5	Japan	D	3i, 2002
Developments in nanotechnology (survey) manufacture	2002	4	USA (West)	D	3i, 2002
Number of publications	1997–1999	3	USA	D	Compano/ Hullmann, 2002
Number of patents (EPO & PCT)	1991–1999	2	USA	D	Compano/ Hullmann, 2002
Number of publications (SCI)	1996–2001	3	USA	D	TAB, 2003
Growth rate of publications	1996–2000	13	South Korea	Netherlands	TAB, 2003
Number of patents (WPINDEX and PATDPA)	1996–2001	2	USA	D	TAB, 2003
Growth rate of patents	1996–2000	4	Canada	D	TAB, 2003
Number of publications for nanotechnology in the field of life science (SCI)	1996–2000	2	USA	D	TAB, 2003
Number of patents for nanotechnology in the field of life science (WPINDEX and PATDPA)	1996–2000	2	USA	D	TAB, 2003
Leading nanotechnology countries (survey) nanomaterials	2003	2	USA	D	Fecht et al., 2003
Leading nanotechnology countries (survey) nanotools	2003	2	USA	D	Fecht et al., 2003
Leading nanotechnology countries (survey) nanodevices	2003	3	USA	D	Fecht et al., 2003
Leading nanotechnology countries (survey) nanobiotech	2003	2	USA	D	Fecht et al., 2003

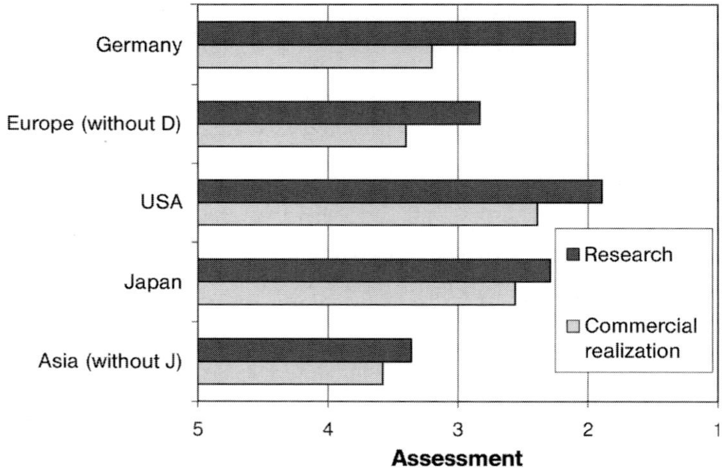

Figure 7.21 Assessment of the international position of nanotechnology with regard to research and realization in the own field of operation. (The results are taken from the paper-based company survey.)

however, a discrepancy in Germany's position was revealed (see Fig. 7.21). While in research and development, Asia, Japan and Europe (without Germany) lag behind the USA and Germany, the companies questioned considered the conversion of nanotechnology into products to be better in the USA and Japan than in Germany.

This allows conclusions to be drawn about the development of the diffusion of nanotechnology in the USA, Japan and Europe. Here, a typical pattern seems to reappear. While research in Germany is seen second in the ranking order, German companies seem to be outstripped again in the commercial conversion of research results into products by faster acting companies in the USA and Japan.

This result, however, is only a subjective evaluation of companies in Germany. For an objective assessment of the success in the commercial realization of nanotechnology, objectifiable measuring parameters would have to be applied.

7.6.2
Development of an Objective Nanotechnology Indicator

To be able to measure progress in the fields of economic and social policy, different indicators have proved their worth as basis for control and report systems. With regard to high-technology fields, only scientific and technological indicators are found in literature, which allow research activities to be assessed on the basis of the number of scientific publications or patents. Technology-related indicators informing about the efficiency and innovative power of companies in a special technological field do not exist. The same goes for the development of employment figures that may serve as a yardstick for the economic realization of a high-technology sector.

For a technology and innovation-related analysis of companies in a special field of application such as nanotechnology, numerous parameters may be of importance, depending on the approach.

Giving priority to the examination of the R&D-activities and the results produced, the following aspects are important for example:
- Extent of R&D expenditure
- Quality of R&D equipment (facilities,...)
- Cooperations
- License portfolio

Analyses of that kind can also be based on the staff size, the changes in and the quality of the R&D-personnel. Parameters for this can be:
- Number of employees in R&D
- Qualification
- Fluctuation/staff turnover rate (in the company)

it is difficult to quantify the parameters mentioned before because they are complex. In order to avoid this inadequacy and to gather statistically valid data, the indicator approach is often applied. An indicator can be interpreted as a proxy parameter of informative value for the actually examined phenomenon. Usually, the application of an indicator is accompanied by a reduction of complexity, since for the evaluation of the facts only one variable is used.

Apart from the individual examinations of company-related data, further data can be used for the examination of the development of company groups, technological fields or lines of industry.

These data aggregate a large number of identical (company) parameters that serve as an indicator for a certain fact. Here, the current availability of the data, their clarity and their comparability inside and outside the scope of analysis are important. For these data, information about the development of employment figures of enterprises on the whole or location-related is worth consideration, as well as the development of their sales, their profits or their investment activities.

In the following, a suggestion for an indicator for the time-related progression of employment figures in the field of nanotechnology is being developed – in consideration of the classification according to F1, F2 or F3 companies. With a comparable sample in other countries (e.g. the USA), a direct "benchmarking" of the manpower development in different countries is possible. Since a relatively high number of companies of different plant sizes are the basis, the sample should have a representative character. This approach is based on the determination and analysis of convincing parameters (indicators) for German nanotechnology companies. These are available through the analysis of existing databases (Markus database etc.) or due to periodical source data acquisition. The use of generally accessible databases has the advantage of being independent from the willingness of individual companies to give information.

Employment indicator

An employment indicator allows a statement about the employment effects and the manpower development of nanotechnology; however, it does not permit conclusions with regard to Germany's economic capability in the field of nanotechnology.

Especially the practicability of data acquisition as well as the physical and subject-related delimitation is important for the development of informative indicators. So employment figures of nanotechnology companies, for example, can be determined in an easy and objective way, whereas the determination of the number of employees in these companies, directly or indirectly dependent on nanotechnology, as well as the part of them working in Germany is expensive and usually based on partly subjective company data. Indirect employment effects on the component supplying industry as well as substitution effects are not taken into consideration.

Nevertheless, a system of specifically selected nanotechnology-indicators is always of informative value, if the determination occurs at regular intervals and over a longer period of time, thus revealing possible trends. Just as useful as a time comparison, can be the comparison with equivalent indicators in other fields of technology, such as biotechnology, or even a "benchmarking" with comparable parameters on an international level.

In any case, an isolated examination of individual parameters involves the risk of paramount, industry-specific or overall economic developments distorting the facts and leading to misinterpretations. Therefore it is indispensable to either integrate such parameters into the system or at least to take such influences into account when evaluating nanotechnology indicators and their development.

Employment index

In Fig. 7.22, an example of the development of an employment index for nanotechnology over the past three years is given on the basis of the available information from the Markus database. (For employment trend, see also Chapter 7.5). This index includes 35 German nanotechnology companies. The number of employees in 2000 was standardized to 100, and the relative deviations of the following years were depicted in comparison.[9] The composition according to company size can be followed from Fig. 7.23.

9) For the year 2003, only the development in comparison to the previous years (2000-2002) was taken as basis for the companies of the reduced sample.

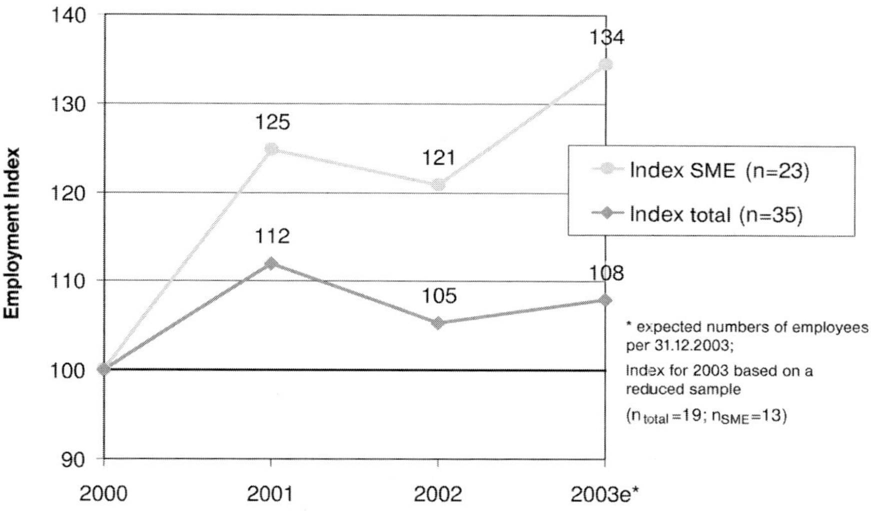

Figure 7.22 Employment index for nanotechnology, separately according to SME and comprehensively for all company sizes (standardized to 100 in the year 2000).

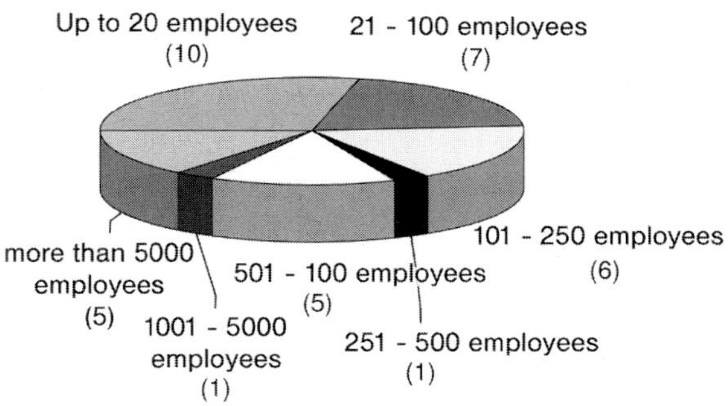

Figure 7.23 Composition of the examined company size.

Figure 7.22 makes clear that after an increase, a steep decline in the employment indicator is to be noticed in the year 2002. Obviously, the overall economic development has a considerable influence on the development of the indicator. The ensuing increase reflects a slight rally in the field of nanotechnology. However, indisputable statements are possible only after a longer period of time.

7.7 Outlook

Evolutionary developments in nanotechnology until 2006

The future prospects until the year 2006 include a short-term period of two to three years. The paper-based company survey, the workshops and the expert interviews took place mainly in the year 2003. In principle, the experts questioned did not expect radical changes in the market environment, so that an evolutionary development rather than a radical and rapid change is to be expected. With regard to the end users of nanotechnology, such as the car industry for example, the periods of time until the introduction of radical innovations exceed the period under review until 2006. Nevertheless, after analysis of the paper-based survey and the workshops, the forecasts for the employment and sales development of the German industry are quite positive for the year 2006, so here a sustainable introduction of nanotechnology as an integral part of the company policies can be assumed.

In the long run, significant changes expected

For a long-term period, however, drastic changes through nanotechnology are expected, partly based on visionary approaches, such as set out in the book "*Unbounding The Future: The Nanotechnology Revolution*" by Eric Drexler, published in 1991. Drexler uses the argument that "engineering" on a molecular level allows a precise and economic control of the basic structure of the matter. Consequently, sheets of a material with a thickness of a few molecules could be manufactured with a hardness of a diamond. Or small microbe-sized machines could be designed to break down toxic waste products, to eliminate plagues or to control virus diseases. The human body could possibly be injected with programmed "nanodevices" that would be delivered to the wounded skin cell for wound healing, while others in turn, applied on a wall, could form display units – as a kind of video wallpaper.

In the current stage of nanotechnology, science and fiction are close together, a fact that occasionally led to empty promises as well as to unrealistic hopes, partly encouraged by the often unsound reporting of the media. Another factor is the still limited physical comprehension of the fundamental mechanisms and effects that sometimes led to technically impracticable promises.

In order to differentiate science from fiction with regard to future products and processes of nanotechnology, a survey was carried out (Fecht et al., 2003) questioning experts about their assessment of the future influence of nanotechnology on specific industrial fields of application. The results shown in Fig. 7.24 are based on the assessment of nearly 50 leading experts at home and abroad.

For the year 2015, the interviewees expect an influence of nanotechnology on nearly all industrial fields. The fields expected to be influenced most by nanotechnology are materials, life sciences and

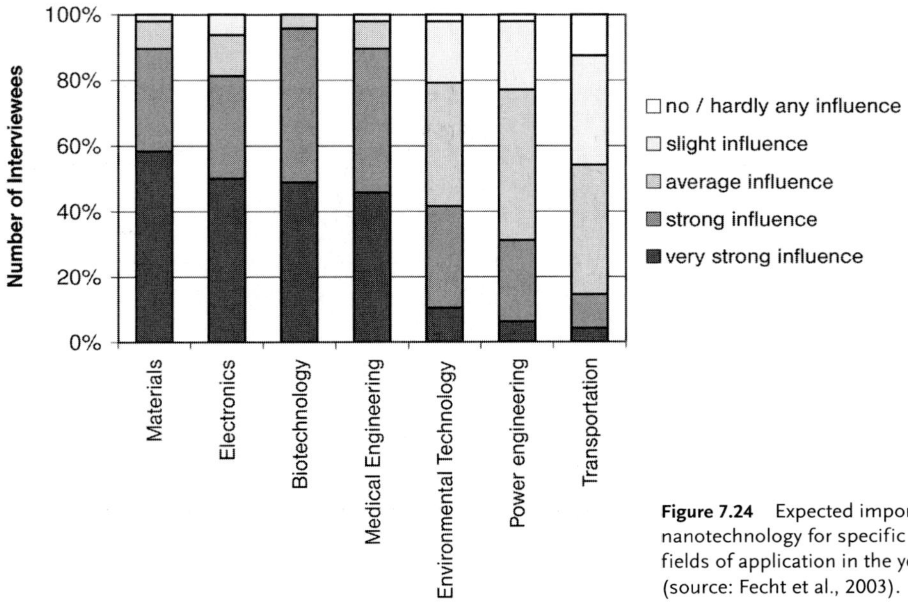

Figure 7.24 Expected importance of nanotechnology for specific industrial fields of application in the year 2015 (source: Fecht et al., 2003).

electronics. Then, at some distance, the fields of environmental technology, power engineering and transportation follow. This assessment of future developments shows an excellent coincidence with the current position of nanotechnology companies in Germany.

Inter alia, the survey includes an evaluation regarding the periods of time to be expected for the commercial availability of socially, economically and technologically important applications and processes. This included:

- the use of self-organization principles for mass production;
- the use of self-replication techniques for mass production;
- products based on nano(bio)technology for the curing of diseases like cancer, Parkinson's disease or Aids;
- products based on nano(bio)technology for the repair of DNA-defects;
- miniaturized machines with the size of molecules (molecular-sized machines);
- nanoelectronics on the basis of new materials and technologies.

The majority of the experts questioned considered a successful commercial realization of the products or technologies mentioned to be unrealistic within the next ten years. Only with regard to the utilization of self-organizing principles as a possibility for mass production, nearly two thirds of those questioned expected commercial availability of such processes before 2015.

In the year 2015, almost every industrial field influenced by nanotechnology

While all experts unanimously consider the use of nanotechnology to be probable in connection with the issues of self-organization, DNA-defects and nanoelectronics sooner or later, 13% regard both self-replication and the future application of "nano-machines" as completely improbable. Only 6% doubt the possibility of overcoming diseases like cancer, Parkinson's disease or aids with the help of nanotechnology.

In future, diffusion of nanotechnology into nearly all industrial fields is expected. It can be assumed that this is an evolutionary development, and the significant commercial breakthrough of nanotechnology, e.g. in medicine or electronics, will not happen immediately, but will still take some time. In these fields, internationally regarded as being influenced most by nanotechnology in future, Germany has an excellent starting position due to its existing industrial power.

7.8
Market Assessment at a Glance

Germany has a good economic starting position in most of the different fields of nanotechnology. Nevertheless, the individual influencing factors on the current stage of nanotechnology in Germany are complex and heterogeneous. Therefore, the essential insights of Chapter 7 are summarized again:

"World market" nanotechnology

- Comparison of different market forecasts for the world market for nanotechnology from different sources shows that the leverage effect of nanotechnology is influencing a world market of about 100 billion EUR. Here it should be recognized that the market volumes of different value-added depths have been added. On average, market forecasts predict an exponential increase over the next ten years. An examination of the individual market shares attributable to Germany is not simple. It is universally accepted that nanotechnology will play a positive role in the future development of German enterprises.

Nanotechnology is no "hype"

- The enterprises polled within the framework of this study disagree with the statement that nanotechnology only represents a new experimental field, just as they reject, in a more moderate way, the statement that through nanotechnology technological competence would be extended. This result clearly refutes the popular opinion that nanotechnology is simply a "hype".
- The results of the company survey showed that the field of chemistry (including materials) clearly heads the list of nanotechnology companies and applications in Germany (judging by the number of companies, the frequency of already existing nanotechnological

products and their sales potential until the year 2006), followed by life sciences (medical technology/health) and I&C. The majority of the competitors in chemistry are regarded to be in Germany and the USA.

Chemistry at the top in Germany

- In the international competition, research in Germany is seen in second place worldwide (after the USA). Regarding the commercial conversion of research into products, German companies are surpassed by the faster acting companies in the USA and Japan, however, they are still among the world leaders, as the results of the company survey showed.

Commercial realization in Germany still to be improved

- The most serious obstacles to innovation in Germany are high investment costs, equity and debt capital and insufficient financial funding. The ranking order of the obstacles mentioned leads to the conclusion that the development of new products or processes in the field of nanotechnology requires considerable investment that cannot be financed by equity capital alone. The development of markets with the help of nanotechnology entails sizeable investments, too, that cannot be raised by industry alone without problems. Limited market knowledge and networks of cooperation that are as yet undeveloped, especially in lines of industry not yet strongly penetrated by nanotechnology, represent barriers for innovation velocity and diffusion of new fields of applications. Cooperation between public finance and nanotechnology companies also represents a significant challenge for the future, a problem not yet solved especially in Germany.

Obstacles to innovation

- The withholding of investments of the venture-capital business in start-up companies has an extremely negative effect on business formations in the German nanotechnology sphere at present. The classical bank-financing of business start-ups has become more and more difficult over recent years, as the German banking landscape itself undergoes a crisis.

Classical bank-financing is difficult

- Assessments regarding the employment growth in nanotechnology suggest an increase of at least 10 000 to 15 000 jobs to be expected until 2006, triggered off by the about 450 companies alone active in nanotechnology in Germany. Even today, approximately 20 000–32 000 (conservative estimate) and 114 000 jobs (optimistic estimate), direct or indirectly dependent on nanotechnology, are to be found in a currently not precisely quantifiable field in Germany.

Employment growth of 10 000 to 15 000 jobs

- For the year 2015 almost all fields of industry are expected to be affected by nanotechnology. As was to be expected, from the international point of view chemistry, life sciences and electronics are the fields influenced most by nanotechnology. Then, at some distance the fields of environmental technology and transporting (inter alia, car manufacture) are following.

Strong future influence of nanotechnology on industrial fields

Table 7.7 SWOT-Analysis of nanotechnology in Germany.

Strengths	Weaknesses
There is a high proportion of German small-sized enterprises active in nanotechnology	The number of medium-sized companies active in nanotechnology is relatively low
Germany disposes of a pronounced strength in the fields of chemistry/ materials in nanotechnology	Partly weak commitment of traditionally strong industrial fields (e.g. mecha nical engineering) to nanotechnology
In the international comparison Germany shows a considerable strength in nanotechnological research	
Opportunities	**Threats**
Intensified commercial realization of partly excellent results of research is required	Financing of nanotechnological acti vities is regarded as one of the greatest challenges (and innovation barriers)
	Suitably qualified staff and adequate cooperative partners are also considered to be a challenge

SWOT-analysis

For a summarizing situation description of complex facts, the SWOT-analysis has proved its worth. The SWOT-analysis set out in Table 7.7 takes up examples of dominant strengths and weaknesses of nanotechnology in Germany. A SWOT-analysis is an analysis of strengths und weaknesses, i.e. the evaluation of factors that may be influenced in Germany itself, and an analysis of opportunities and threats, i.e. an evaluation of factors partly having global effects.

The SWOT-analysis underlines the most striking statements of the paper-based company survey and the results of the workshops. The financing of nanotechnology companies, mentioned under "Threats", especially in connection with the withholding of investments of "venture-capital enterprises" in start-up companies turns out to be a problem.

Germany has an excellent starting position

The summary reveals that Germany has an excellent starting position for the economic realization of nanotechnological activities. The excellence of research, however, is not reflected completely in the economic realization. Here, the USA and Japan have been superior to Germany so far. It must also be taken into account that investments and public funding in the field of nanotechnology have increased considerably all over the world, a fact partly attributable to the forecast very high market volumes. In future, a stronger international competition in nanotechnology is to be expected, in which players from outside the USA, Japan and Germany will appear increasingly.

International competition increases

8
Summary and Conclusions

Norbert Malanowski and Axel Zweck

The present survey was intended to be a realistic assessment of market volume and relevance of nanotechnology both for Germany itself and in an international context. With regard to the acquisition and analysis of data, a comprehensive methodological mixture has proved its worth and helps to avoid the deficits of the prevailing qualitative (expert interviews, literature analysis, Delphi-workshops) and quantitative (paper-based company survey and patent analysis) methods for the determination of the economic potential of nanotechnology. In this way it is possible to survey the individual results not only exclusively or even isolated from each other within the framework of the special method applied, but to deal with them within the context of all the results. For the evaluation of the results it was particularly useful that in industry-specific workshops, (provisional) results could be critically examined and evaluated by acknowledged experts from industry, science and finance . With this, even elements of participative innovation and technology analysis were brought into this survey as impetus for a steady and constructive discussion with the stakeholders.

Mixture of methods proved its worth

It is certainly not the purpose of the last chapter to repeat all the main results. Such an overview of results can be found both on the first pages of this study, and in the summaries at the end of the individual chapters. Here, we would only like to draw attention to the fact that:

- Nanotechnology is one of the *key technologies* of the 21st century. These days, considerable turnover is achieved with products that are only feasible by the existence of nanotechnology. With the economic breakthrough of nanotechnology, this turnover should increase significantly in the future.

  "Nanoproducts" even today

- Germany is very well positioned with regard to its *patent situation* in nanotechnology – both in nanotechnology as a whole and in the quantitatively most important section of chemistry. Germany's position appears to be particularly good regarding the especially valuable patents. In none of the Lead Markets (chemistry, car manufacture, and optics) that were surveyed in the patent analysis is Germany lagging seriously behind the USA or Japan.

  Germany very well positioned with patents

Growth Market Nanotechnology. Edited by N. Malanowski, T. Heimer, W. Luther, and M. Werner

ISBN: 3-527-31457-1

- It is not possible to evaluate the "nanotechnological world market" on the basis of the figures mentioned in publicly available surveys. This is because market data is only available for part of the nanotechnological products and thus the lists are incomplete; market forecasts partly refer to different time horizons; nanotechnological products are named twice in two or more sections (e.g. application of nano-basic-products/components in end products of different lines of industry) and products from different stages of the value-added chain are included in the survey (basic products, intermediate products, end products etc.).

Reservations about the term "world market nanotechnology"

- The withholding of investments of the *venture-capital* business in start-up companies has an extremely negative effect on business formations in the German technological sphere at present. The classical bank-financing of business start-ups has become more and more difficult in recent years, as the German banking landscape itself is undergoing a crisis.

VC and classical bank-financing difficult at present

- Germany currently has an excellent starting position for the *economic realization* of its nanotechnological activities. Its excellence in research, however, is not completely reflected by economic realization. In this area the USA and Japan have been superior to Germany so far.

Germany has an excellent starting position

- Investment and public funding in the field of nanotechnology have increased considerably all over the world, a fact partly attributable to the very high market volumes predicted. Therefore, even stronger *international competition* with regard to nanotechnology can be expected in future.

Competition becomes more severe

Apart from the extremely necessary and regular surveys on the market potential of nanotechnology, the following activities within the context of the innovation and technology analysis would be advisable to enable the exploitation of the largest share possible of the forecast huge market potential of nanotechnology in future markets in Germany:

Exploiting future markets

- The establishing of a *nanotechnological indicator*, for example for the employment trend in nanotechnology, in order to have a measurable reference value in hand, inter alia for the efficiency of governmental funding programs, as well as a "benchmarking" with equivalent indicators of other technological fields, such as biotechnology or with comparable parameters in the international comparison (see Chapter 7.5.2 of the survey). In this way, the time development of the economic importance of nanotechnology can be followed. For future studies, subdivision of nanotechnology companies into producers, users etc. could be practical.

- The development of an *internet-based platform* for German nanotechnology companies, to facilitate the marketing of nanotechnological products abroad and the expansion of business activities in international growth markets in Asia.
- Early *studies on possible show-stoppers,* such as the toxicity of certain nanomaterials, to minimize the risk of bad investments or market barriers owing to lack of consumer acceptance.

References

3i: "Nanotechnology – Size matters", A 3i white paper in association with the Economist Intelligence Unit and the Institute of Nanotechnology, London (2002)

Alemann, U. v.: "*Einführung in die Methoden der Politikwissenschaft*", Opladen (1995)

Aschenbrenner: "NTC Nano tech coatings – Wissenschaft in Unternehmen: Ohne Chrom geht's auch, Korrosionsschutz durch neuartige Beschichtung", in: *Spektrum der Wissenschaft* (2003)

Aspen Systems: "Aspen Systems takes a giant step toward commercialization of Aerogels", press release (2001)

BASF/Distler: "Nanoteilchen in Megatonnen: Vielfältige Anwendungen für Polymerdispersionen", in: BASF press release (2002)

Bachmann, G.: "Innovationsschub aus dem Nanokosmos – Technologieanalyse", VDI Technology Center, *Future Technologies* **28**, Düsseldorf (2001)

BCC: "Biomedical Applications of Nanoscale Devices: Commercial Opportunities", Conference proceedings, Nanotech and Biotech Convergence, BCC, Stanford (2003)

BCC: "Global Nanotechnology Market to Reach $29 Billion by 2008", press release of 03.02.2004, Business Communication Company, (http://www.bccresearch.com/editors/RGB-290.html) (2004)

BCC: "Ion Implantation and MBE Technologies to grow at 25 per cent and 33 per cent through 2006", press release (http://www.bccresearch.com/editors/RGB-201B.html) (2002)

BCC: "Opportunities in Nanostructured Materials: Biomedical, Pharmaceutical & Cosmetic", Business Communication Company, press release (2002) (http://www.bccresearch.com/editors/RGB-186F.html) (2002)

Becker: "Markt- und Technologietrends bei Flüssigkristalldisplays", lecture Merck KgaA (2003), (http://pb.merck.de/servlet/PB/show/1280810/TAE_2003_Manuskript.pdf) (2003)

Beckmann, M., Lenz, P.: "*Profitieren von Nanotechnologie. Investment der Zukunft*", Munich (2002)

Beise, M.: "Identifizierung von "Lead-Markets": Ein neuer Ansatz einer globalen Produkteinführungsstrategie", in: BMWI: "Die innovative Gesellschaft – Nachfrage für die Lead-Märkte von morgen", results of the convention in Berlin (2002)

BMBF: "*Zur technologischen Leistungsfähigkeit Deutschlands*", Bonn (2001)

Growth Market Nanotechnology. Edited by N. Malanowski, T. Heimer, W. Luther, and M. Werner

ISBN: 3-527-31457-1

BMBF (Federal Ministry of Education and Research), Department for Public Relations: "IT-Forschung 2006, Förderprogramm Informations- und Kommunikationstechnik", Bonn (2002)

BMBF: "*Nanotechnologie in Deutschland – Standortbestimmung*", Bonn (2002)

BMBF: "*Bericht zur technologischen Leistungsfähigkeit Deutschlands*", Bonn (2002)

BMBF: "*Nanotechnologie erobert Märkte – Deutsche Zukunftsoffensive für Nanotechnologie*", Bonn (2004)

BMWi: "Die innovative Gesellschaft – Nachfrage für die Lead-Märkte von morgen", results of convention, Berlin (2002)

Bosch: "*Harte Schichten für hohe Beanspruchungen*", Bosch Research Info (2001)

Braun, T., Schubert, A., Zsindely, S.: "Nanoscience and Nanotechnology on the Balance", in: *Scientometrics*, **38** (2) (1997)

Bruchmann, B.: "Baummoleküle im Nanomass-stab: Dendrimere für neue Drucksysteme und Autolacke", BASF press release, Mannheim (2002)

Carillo, D.: "Nanosensors' Niche in Nanotechnology", in: Cepmagazin-online (www.cepmagazine.org/pdf/110333.pdf) (2003)

Cebulla, E.; Malanowski, N.; Zweck, A.: Innovation Accompaniment Nanotechnology: University programs in the field of nanotechnology, Future Technologies Consulting Volume 57; Editor: Future Technologies Consulting, VDI Technology Center GmbH on behalf of the BMBF (2006)

Chilcott, J., Jones, A., Mitchell, M.: "*Nanotechnology: Commercial Opportunity*", Evolution Capital Ltd., London (2001).

Compano, R., Hullmann, A.: "Forecasting the development of nanotechnology with the help of science and technology indicators", in: *Nanotechnology* **13** (2002)

Compano, R.: "Technology Roadmap for Nanoelectronics", IST Programme Future and Emerging Technologies, Brussels (2001)

Critchlow, Dale L.: "MOSFET Scaling – The Driver of VLSI Technology," *IEEE Proc.* **87**(4) (1999)

Dennard, R. H., Gaensslen, F. H., Kuhn, L., Yu, H. N.: "Design of Micron MOS Switching Devices", presented at the IEEE International Electron Devices Meeting (1972)

Deschamps, J.-P.: "Managing the Innovation Process", International Institute for Management Development, Lausanne (2000)

Distler, D.: "Nanoteilchen in Megatonnen: Vielfältige Anwendungen für Polymerdispersionen", BASF press release, Mannheim (2002)

Ebenau, A.: "Wirtschaftliche Perspektiven der Nanotechnologie: Enorme Märkte für kleinste Teilchen", Workshop Nanotechnologie in der Chemie – Experience meets Vision, Mannheim (2002)

Fecht, H.-J., Ilgner, J., Köhler, T., Mietke, S., Werner, M.: "Nanotechnology Market and Company Report – Finding Hidden Pearls", WMtech Center of Excellence Micro and Nanomaterials, University of Ulm (2003)

FHG-INT: "Anwendungspotenziale nanotechnologiebasierter Materialien", Expert's report in cooperation with VDI TZ on behalf of the TAB, Fraunhofer-Institut für Naturwissenschaftlich-Technische Trendanalysen, Euskirchen (2003)

Gordon, N.: "Job Opportunities in Nanotechnology", William M. Tatham Centre for Co-operative Education & Career Services, University of Waterloo (Canada), May 21 (2003)

Front Line Strategic Consulting: Nanobiotechnology, press release (www.smalltimes.com/document_display.cfm?document_id=6060) (2003)

Frost&Sullivan: Stevenson, R., 2002: OLEDs set to glow. e-zine chemSoc; (http://www.chemsoc.org/chembytes/ezine/2003/stevenson_jan03.htm)

Frost&Sullivan: Carillo, D.: Nanosensors' Niche in Nanotechnology. In: Cepmagazin-online, http://www.cepmagazine.org/pdf/110333.pdf (2003)

Haas, K.-H. et al.: "Produktion von Nanomaterialien", final report on the BMBF-Project (FKZ 02PH 2107), carried out by FHG-ISC and FHG-ISI, Karlsruhe (2003)

Heimer, T., Werner, M.: "Die Zukunft der Mikrosystemtechnik – Quo vadis MST?", Ergebnisse der ex-ante Evaluation des Förderkonzepts Mikrosystemtechnik 2000+, Wiley-VCH, Weinheim (2004)

Hellebrand, O., Kaube, G.: "*Lizenzsätze für technische Erfindungen*", 2nd edn, Carl Heymanns Verlag, Cologne (2001)

Hoffknecht, A., Hoffschulz, H.: "Technologiefrüherkennung, Ergebnisse des Fachgesprächs Molekularelektronik", *Future Technologies*, **39**, Düsseldorf (2002)

Hoffknecht, A.: "Technologiefrüherkennung, Elektronik der Zukunft, Mini-Delphi-Studie", *Future Technologies* **46**, Düsseldorf (2003)

Hohenberger, W.: "*Submikron und Nano – Revolution bei Füllstoffen und Compounds; Kunststoffe*", Munich (2000)

Horn, D. und Rieger, J.: "Organische Nanopartikel in wässriger Phase – Theorie, Experiment und Anwendung", *Angewandte Chemie* **113** (2001)

Huang, Z. et al.: "Longitudinal Patent Analysis for Nanoscale Science and Engineering: Country, Institution and Technology Field", in: *J. Nanoparticle Research*, **5** p333–363 (2003)

Hullmann, A.: "*Internationaler Wissenstransfer und technischer Wandel – Bedeutung, Einflussfaktoren und Ausblick auf technologiepolitische Implikationen am Beispiel der Nanotechnologie in Deutschland*", Physica-Verlag (Springer), Heidelberg (2001)

ITRS (International Technology Roadmap for semiconductors): (http://public.itrs.net/Files/2003ITRS/Home2003.htm) (2003)

ITRS (International Technology Roadmap for semiconductors): (http://public.itrs.net/files/1999_SIA_Roadmap/) (1999)

ITRS (International Technology Roadmap for semiconductors): (http://public.itrs.net/Files/2001ITRS/Home.htm) (2001)

ITRS (International Technology Roadmap for semiconductors): Update, (http://public.itrs.net/Files/2002Update/2002Update.pdf) (2002)

Jankowski, P.: "*Mikrosystemtechnik und Nanotechnologie – Schlüsseltechnologien für Deutschland*", Sal. Oppenheim, Cologne (2001).

Langenfeld, S., Friedrich, H., Meyer, C.: "Adjusted Functional Surfaces by Nanotechnology for Automotive Applications", conference report, Materials Week 2001, International Congress on Advanced Materials, their Processes and Applications; Munich (2001)

Löfken, J. L., Mayr, P. M.: "Operation Schwarzer Schnee", in: *Technology Review*, German issue (2003)

Managermagazin: "Wo Deutschland Spitze ist", Vol. **9** (2002)

Mann, C. C.: "The End of Moore's Law", *Technology Review* May/June p 43–48 (2000)

Marinova, D., McAleer, M.: "Nanotechnology strength indicators: International rankings based on US patents", in: *Nanotechnology* **14** R1–R7 (2002)

Masaki et al.: "Creep-strengthening of steel at high temperatures, using nanosized carbonitride dispersions", in: Nature **424**, 294–296 (2003)

Materials World: "Nanotechnology toughens up steel with tiny precipitates", *Materials World* **9**(7) (2001)

Meyer, M. S.: "Patent Citation Analysis in a Novel Field of Technology: An Exploration of nano-science and nanotechnology", *Scientometrics* **51**(1) (2001)

Meyer, M., Persson, O., Power, Y., Nanotechnology Expert Group and Eurotech Data: "Mapping Excellence in Nanotechnology", preparatory study, (http://europa.eu.int/comm/research/era/mapping-excellence.html) (2001)

Meyer, M., Persson, O.: "Nanotechnology – Interdisciplinarity, Patterns of Collaboration and Differences in Application", in: *Scientometrics* **42**(2) (1998)

Mitsubishi Research Institute 2002: "Promoting Japanese-style Nanotechnology Enterprises" presentation of S. Kamei at the 1st Jetro International Nanotechnology Symposium on 26th of February in Chicago, US

Mitsui & Co. Ltd.: Press release (2002)

Moore, Gordon E.: *Electronics* **38**(8) 114–117 (1965)

Moore, Gordon E.: http://www.semichips.org/downloads/Moore.pdf (2002)

Müller, U.: "Nanowürfel als Wasserstoffspeicher", BASF press release, Mannheim (2002)

Normile, D.: "Court tells Nichia to Pay Blue LED Inventor 180 Million USD", *Science* **303**(5659) (2004)

Normile, D.: "Blue LED Inventor Sues Former Company", *Science* **293**(1575) (2001)

OECD: "Patent Manual 1994 – Using Patent Data as Science and Technology Indicators", Paris (http://www.oecd.org/dataoecd/33/62/2095942.pdf) (1994)

Pitkethly, R.: "The Valuation of Patents", background article for the first meeting of the "High Level Task Force on Valuation and Capitalization of Intellectual Assets" (2002)

Platinum Association: "PGMs Enabling a New Data Storage Technology", *Newsletter International Platinum Association* (2003)

Reuters: "Degussa investigations into alleged price-fixing in the carbon black industry", press release (2002)

Rittner, M. N.: "Nanoparticles – What's Now, What's Next?", in: Nano-Technology [Chemical Engineering progress 99(11)] (2003)

Rittner, M. N.: "Market Analysis of Nanostructured Materials", *Am Ceram Soc Bull* **81**(3) (2002)

Roco, M. C.: "The Vision and Action Plan of the National Nanotechnology Initiative", http://www.cr.org/publications/MSM (2002)

Roco, M. C.: "International perspective on government nanotechnology funding in 2005", in: *Journal of Nanoparticle Research*, **7**(6) (2005)

Rödiger, M., Fuchs, Patrick, Müller, H.: "Zukünftiger Wachstumsmotor in der Chemieindustrie", in: Im Fokus – Nanotechnologie in der Chemie, DG-Bank, GZ-Bank AG (Editor.), Frankfurt (2001)

Rügheimer, U.; Schiller, R.: Von Zwergen und Mottenaugen – Audi zeigt Nanotechnologien auf der Hannover-Messe; Audi AG, Kommunikation Produkt und Technik, Ingolstadt (2002)

S.I.A. (Semiconductor Industry Association): Roadmap, (http://techreports.larc.nasa.gov/ltrs/PDF/1998/cr/NASA-98-cr207652appE.pdf) (1995)

Sepeur-Zeitz: "Nanotechnologie im Automobilbau", information booklet, VW AG, (2003)

Service, R.: "Optical Lithographie goes to Extremes – and beyond", in: *Science* **293**, (2001)

Small Times: "Veeco came, saw, acquired majority of the AFM Market", www.smalltimes.com (2002)

Spectaris: "Branchenbericht 2003", Cologne (2003)

SRI: "Nanoscale chemicals and materials: An overview on technology, products and applications", in: SRI-International Report, Speciality Chemicals: Nanotechnology (2002)

Stevenson, R: "OLEDs set to glow", e-zine chemSoc. (www.chemsoc.org/chembytes/ezine/2003/stevenson_jan03.htm) (2003)

TAB: "TA-Projekt Nanotechnologie Endbericht", Arbeitsbericht Nr. 92, TAB Büro für Technikfolgen-Abschätzung beim Deutschen Bundestag (2003)

VCI: "Chemiewirtschaft in Zahlen", issue 2003, Verband der Chemischen Industrie, Frankfurt/ Main (2003)

VDA: Annual Report 2003, Association of the Car Industry, Frankfurt/Main (2003)

VDI-News, Issue published March 28 (2003)

VDI TZ Expert Workshop (2003)

Verbeek, A., Callaert, J.: "Linking Science to Technology – Bibliographic References in: Patents", Vol. 9, Detailed analysis of the science technology in the field of Nanotechnology, EUR20492, Office for Official Publications of the European Communities, Luxemburg (ftp://ftp.cordis.lu/pub/indicators/docs/ind_report_kul9.pdf) (2002)

Wacker Silicones: News (www.wacker.com) (2003)

Wagner, V., Wechsler, D.: "Nanobiotechnologie II: Anwendungen in der Medizin und Pharmazie", *Zukünftige Technologien*, **50**, Düsseldorf (2004)

Waser, R.: "*Nanoelectronics and Information Technology. Advanced Electronic Materials and Novel Devices*", Wiley-VCH, Weinheim (2003)

Wechsung, R., Fatatry, A. E., Götz, F., Illing, M. Lawes, R. A., Tschulena, G., van de Weijer, F., Wicht, H., Zinner, H. (2002): "Market analysis for microsystems II – 2000-2005", NEXUS Task Force Report (2002)

Werner, M., Mietke, S., Ilgner, J., Köhler, T.: "Nanotechnologie – Zukunftsperspektiven und Marktpotenziale", in: *Laborpraxis* (2003)

Werner, M.: "Markets and Industry-Relevant Research Directions for Nanomaterials and Their Applications", in: *IUMRS Facets*, **2**(3) (2003)

Wollmann, H.: "Politikevaluierung, Evaluierungsforschung", in: "*Kleines Lexikon der Politik*", Nohlen, D. (Ed.), Munich (2001)

Wurzer, A. J.: "Patentbewertung – Technologische und wirtschaftliche Aspekte", workshop discussion of the Zentrum für Gewerblichen Rechtsschutz of the University of Düsseldorf, (http://www.gewrs.de/files/wg_2002_0415_patentbewertung.pdf) (2002)

Appendix 1

Growth Market Nanotechnology. Edited by N. Malanowski, T. Heimer, W. Luther, and M. Werner

ISBN: 3-527-31457-1

VDI-Technologiezentrum

VDI

BMBF Project "The Economic Potential of Nanotechnology"

- Discussion manual for expert interviews -

- How would you classify your unit of organization (e.g. combined group, subsidiary, start-up, large-scale enterprise, SME, spin-off, department/sector, task force etc.)?
- When was this unit of organization founded? How many employees are there currently? How many employees does the company have altogether?
- How many jobs depend on the application of nanotechnology?
- Please state the development of the following operational parameters during the past five years: turnover, export, R&D-expenses.
- Since when do you concern yourself with aspects of nanotechnology – observation, development, use, other?
- For which of the following fields of application is nanotechnology of importance for your company today? (State fields of application or present list)
- Where is the main emphasis put on with regard to your business activity? (Adaptation of possible answers according to the present classification of industrial segments)
- To what extent is your company engaged in R&D in the field of nanotechnology?
- On which fields of application are your efforts of research focused?
- To which stage of the value-added chain do your products belong? – Basic materials, primary product, end product, (services)?
- How are your customers distributed geographically?
- What turnover did your company make over the last year (2001) a) altogether, b) with products depending on nanotechnology and produced by yourself (explanation required)?

- What savings potentials are to be expected for your company (explain: e.g. production, transport, energy etc.)?
- What is the share of supplied nanotechnological products in the total turnover? 1996/2001?
- Which nanotechnological fields of application affect the products manufactured by you? (Adaptation of possible answers according to the present classification of industrial segments)
- Which nanotechnological fields of application affect your supplied products? (Adaptation of possible answers according to the present classification of industrial segments.)
- Are there any obstacles impeding the desired production of nanotechnological products? If so, which essential obstacles had and have an impeding effect on the realisation of the desired application of nanotechnology?
- How do you assess the position of your company with regard to research and realization of nanotechnology in an international comparison?
- Do you plan to enhance or to reduce your activities in the field of nanotechnology over the next five years or to carry on as usual?
- Which fields of application will gain importance during the next five years in a) research and b) realization?
- What total turnover do you expect from turnovers arising from products with integrated nanotechnology in 2002? Germany/internationally?
- What market share do you expect for 2006 on the market you are represented on with your products? Germany/internationally?
- What employment trend do you expect due to products depending on nanotechnology up to 2006?

- Which job outlines / qualifications are relevant for jobs in the context of nanotechnology in your field? Are there any modifications required? If so, which?
- Which hopes do you place on public funding of innovations?
- Which hopes do you place on private funding of innovations (e.g. venture capital)?
- Thinking of your expectations regarding the use of nanotechnology about 5 years ago, from today's view have your expectations come true (list)?

Appendix 2

Growth Market Nanotechnology. Edited by N. Malanowski, T. Heimer, W. Luther, and M. Werner

ISBN: 3-527-31457-1

Survey on the „Economic Potential of Nanotechnology"

The questionnaire was completed by

Department ______________________

Function ______________________

Address ______________________

Phone ______________________

E-Mail ______________________

Company stamp

Please return the questionnaire by **February 21, 2003, at the latest** to:
Hochschule für Bankwirtschaft, Dr. H. Sanders, Sonnemannstrasse 9-11, 60314 Frankfurt/Main, Phone: 069-154008 703; FAX: 069-154008 728

According to our estimate, the completion of the questionnaire will require about 20 minutes, depending on the commitment of your company to nanotechnology. This effort shall be worthwhile for you! Therefore we offer you the cost-free delivery of the results of our survey.
Should we send you the results of this survey?

❑ Yes ❑ No

This survey was based on the following ***definition of nanotechnology:***

Here, nanotechnology is interpreted as:

a) All products with at least one functional component with a controlled geometrical size below 100 nanometers in at least one directional dimension rendering physical/chemical or biological effects usable that do not occur above this critical size.

And

b) Analytical and/or process engineering equipment required for the controlled manufacturing, positioning or measurement of the functional components mentioned under a).

All data is treated confidentially according to the Federal Data Protection Act. No data will be passed on to third parties.

I. Questions concerning the entire company

1. When was your company set up?

Year of foundation: ________

2. Is your company economically independent? ❑ Yes ❑ No

If not, what is the share of the largest affiliated company (group, holding etc.) in your company?

❑ less than 25 % ❑ 25 – 50 % ❑ 51 – 99% ❑ 100 %

3. Where is the major shareholder based?

❑ D ❑ EU ❑ foreign countries (without EU)

4. Do venture-capital investors hold an interest in your company? ❑ Yes ❑ No

If so, what is their share?

❑ below 25 % ❑ 25 – 50 % ❑ 51 – 75% ❑ 76 - 100 %

5. Line of industry: Where is the focus of the business activity of your company? (Single reference)

Textile and clothing business	❑	Manufacture of industrial process control plants	❑
Chemical industry	❑	Manufacture of optical equipment	❑
Rubber and plastics business, manufacture	❑	Manufacture of motor vehicles and motor vehicle parts	❑
Glass and ceramics trade	❑	Rail vehicle engineering	❑
Metal production and working	❑	Aircraft and aerospace engineering	❑
Mechanical engineering without machine tools and domestic appliances	❑	Manufacture of electricity distribution and switching devices	❑
Machine tool engineering	❑	Manufacture of telecommunication devices and equipment	❑
Domestic appliances	❑	Manufacture of electronic components	❑
Manufacture of office equipment, DP-appliances and equipment (without medical, measurement and control engineering, optics)	❑	Software houses, DP services	❑
Manufacture of medical engineering equipment	❑	Telecommunications services	❑
Manufacture of measuring, control, navigation instruments and devices and similar things	❑	Scientific research facilities and institutes	❑
Others ______________________	❑	Technical, physical and chemical surveys	❑

6. How is the distribution (in percent of the turnover) of your customers ?

_____ % Germany _____ % Europe (without D) _____ % USA

_____ % Japan _____ % Asia (without J) _____ % Others

7. Which turnover did your company make in the financial year 2001?

❑ up to 2 million. € ❑ 2 to 10 million € ❑ 10 to 50 million € ❑ more than 50 million €

Your expenditure for research and development was ____ % of the last annual turnover.

8. How many employees does your company have?

❑ up to 20 ❑ 21 to 100 ❑ 101 to 250 ❑ 251 to 500

❑ 501 to 1000 ❑ 1001 to 5000 ❑ 5000 and more

II. The importance of nanotechnology for your company/ies/corporate unit/staff unit (e.g. R&D)

9. For whom do you answer the questionnaire in the following?

❑ entire company ❑ corporate unit ❑ staff unit (Strategy-/R&D)

10. Since when have you been dealing with nanotechnological matters?

	before 1990	*1990 to 1995*	*1995 to 2000*	*since 2001*
Investigation of the scene	❑	❑	❑	❑
Own R&D-works	❑	❑	❑	❑
Utilization in product	❑	❑	❑	❑
Others ____________________	❑	❑	❑	❑

11. Which sales share (in % of the turnover of the whole company) is attributable to products in which nanotechnology plays a functional role or that have been manufactured with the help of nanotechnology?

	1996	2001
Turnover	________ %	________ %

12. In which way are you involved in nanotechnology?

❑ ***as manufacturer*** ❑ ***as user of nanotechnology***

If as user, what share (in % of the total company turnover) is attributable to primary products in which nanotechnology plays a functional role or that have been manufactured with the help of nanotechnology supplied to you by other manufacturers (purchasing volume)?

1996 ________ % 2001 ________ %

13. Where do you obtain your complementary nanotechnological products from? (in % of the purchasing volume)

_____ % Germany   _____ % Europe (without D)   _____ % USA

_____ % Japan   _____ % Asia (without J)   _____ % Others

14. How many employees of your entire company depend (directly or indirectly) on nanotechnology?

❑ up to 5   ❑ 6 to 25   ❑ 26 to 100   ❑ 101 to 250   ❑ 250 and more

15. For which of the following fields of application are your nanotechnological products of importance <u>today</u>? (Multiple references possible)

❑ Information and Communication
❑ Medicine / Life Sciences
❑ Cosmetics
❑ Transportation and Traffic
❑ Chemistry / Materials and Process Engineering
❑ Environment (incl. Recycling)
❑ Energy (-supply)
❑ Mechanical Engineering
❑ Measurement Technology
❑ Building Trade
❑ Food / -processing
❑ White and Brown Goods
❑ Others, which?: ____________________

16. How do you assess the importance of nanotechnology (NT) for your company as a whole?

	applicable 1	2	3	4	*not applicable* 5
Our NT know-how is a decisive competition factor today	❑	❑	❑	❑	❑
NT improves our technological competitiveness in our traditional markets	❑	❑	❑	❑	❑
NT helps us to tap into completely new markets	❑	❑	❑	❑	❑
NT rounds off our technological competence, however it will not be developed to a core competence	❑	❑	❑	❑	❑
NT is one of many technological options we are pursuing	❑	❑	❑	❑	❑
NT is a new experimental field for us	❑	❑	❑	❑	❑

17. How do you assess the international level with regard to research and realization of nanotechnology in your field of activity?

Please assess the geographical regions from **1 to 5** (1 = best; 5 = worst) with regard to the current stage of development.

	Level of research	*Level of commercial realization*
Germany	☐	☐
Europe (without D)	☐	☐
USA	☐	☐
Japan	☐	☐
Asia (without J)	☐	☐
Others ____________	☐	☐

18. Where is the headquarter of your major competitor with regard to your nanotechnological products / fields of application?

❑ Germany ❑ Europe (without D) ❑ USA

❑ Japan ❑ Asia (without J) ❑ Others

III. Future prospects

19. Are you planning to intensify your activities in the field of nanotechnology until 2006?

intensify considerably	intensify	unchanged	reduce considerably	abandon completely	I do not know
❑	❑	❑	❑	❑	❑

20. Do you reckon on an increase in labour employment in the field of nanotechnology until 2006?

❑ Yes ❑ No

If so, to what extent

❑ below 10 % ❑ 10 – 20 %

❑ 21 – 50 % ❑ 51 – 75 %

❑ 76 – 100 % ❑ 101 – 200 %

❑ over 200 %

21. **Please tick the box indicating the <u>3 most important future fields of application for your nanotechnological products</u> (in 2006) in the following list** (left-hand box) **and enumerate an order from 1 to 3 according to their importance** (right-hand box). **Here, such fields of application are regarded as important in which you see great economic chances for your company owing to the use of nanotechnology.**

Fields of application (analogous to question 15, here, however, ***in 2006***)	Order *[please insert numbers (1 – 3)]*
❑ Information and Communication	☐
❑ Medicine / Life Sciences	☐
❑ Cosmetics	☐
❑ Transportation and Traffic	☐
❑ Chemistry/Materials and Process Engineering	☐
❑ Environment (incl. Recycling)	☐
❑ Energy (-supply)	☐
❑ Mechanical Engineering	☐
❑ Measurement Technology	☐
❑ Building Trade	☐
❑ Food / -processing	☐
❑ White and Brown Goods	☐
❑ Others, which: ____________	☐

Selection of the **three** most important fields of application

22. **With regard to these 3 fields of application, which functions are of importance for your nanotechnological products (according to the order made up in question 21) ?** (Multiple references possible for each field)

Possible functions	*Field 1*	*Field 2*	*Field 3*
Analytics/ Diagnosis	❑	❑	❑
Medical Therapy / Diagnosis	❑	❑	❑
Surface functionalization	❑	❑	❑
Displays	❑	❑	❑
Energy conversion	❑	❑	❑
Manufacturing equipment	❑	❑	❑
Nanobiological functions	❑	❑	❑
Data processing and storage	❑	❑	❑
Data transfer (Telematics)	❑	❑	❑
Material separation	❑	❑	❑
Sensor technology	❑	❑	❑
Actuators	❑	❑	❑
Material dosage	❑	❑	❑
Optical effects	❑	❑	❑
Filtering of fluids or gases	❑	❑	❑
Protection (against corrosion, dirt etc.)	❑	❑	❑
Improved material properties	❑	❑	❑
Structure building	❑	❑	❑
Design / Fashion / Esthetics	❑	❑	❑
Others, which: ____________	❑	❑	❑
____________	❑	❑	❑
____________	❑	❑	❑

23. Which technologies do you apply for the realization of your nanotechnological products in the 3 fields of application named by you in question 21? (Multiple references possible for each field)

Possible Technologies	*Field 1*	*Field 2*	*Field 3*
Self assembly	❑	❑	❑
Molecular engineering	❑	❑	❑
Biological engineering	❑	❑	❑
Ultra-precision engineering	❑	❑	❑
Separation / filtration methods	❑	❑	❑
Sol-gel processing	❑	❑	❑
Powder processing	❑	❑	❑
Catalysis	❑	❑	❑
Thin film deposition methods	❑	❑	❑
Optical lithography	❑	❑	❑
Particle beam lithography	❑	❑	❑
Nanoprint/-imprint	❑	❑	❑
Microscopy	❑	❑	❑
Metrology	❑	❑	❑
Other Analytical Methods	❑	❑	❑
Modeling and Simulation	❑	❑	❑
Others, which: ________________	❑	❑	❑
________________	❑	❑	❑

24. Which of the products/processes based on nanotechnology are already or will be of great economical importance in the fields of application named by you in question 21? (Multiple references possible)

	Product/Process:	*of economic importance* Today	until 2006	from 2006
in application field 1	________________	❑	❑	❑
	________________	❑	❑	❑
	________________	❑	❑	❑
	Product/Process:			
in application field 2	________________	❑	❑	❑
	________________	❑	❑	❑
	________________	❑	❑	❑
	Product/Process:			
in application field 3	________________	❑	❑	❑
	________________	❑	❑	❑
	________________	❑	❑	❑

IV. The following questions (25-29) refer exclusively to the field of application ranking first in your order defined in question 21

25. Which steps have you already taken to be able to use application field 1 with the help of nanotechnology?

Carrying out real investments	❑
Contacting cooperative partners	❑
Acquisition of know-how through purchase of property rights	❑
Acquisition of know-how through projects for research and cooperation	❑
Appropriation of research staff	❑
Development of an international distribution network	❑

26. In which phase are your works on application field 1 currently?

❑ Research ❑ Development ❑ Prototype ❑ Product ❑ Patent

27. How do you assess the global market in 2006 for the most promising product you have named?

Please name this product again (comp. question 24): ____________________

I assess the **volume of the world market** at

❑ less than 50 million € ❑ 50 – 250 million €

❑ 250 - 500 million € ❑ 500 – 1000 million €

❑ more than 1000 million € __________ million €

namely

I assess **Germany's share** in the world market for this product in 2006 at

❑ less than 1 % ❑ 1 – 10 % ❑ 11 – 25 % ❑ over 25 %

I assess the **world market share of our company** for this product in 2006 at

❑ less than 1 % ❑ 1 – 10 % ❑ 11 – 25 % ❑ over 25 %

28. What are your market assessments based on? (Multiple references possible)

Own market investigations ❑

Customer survey ❑

Market surveys ❑

Press information ❑

Statements of competitors ❑

Others ____________________ ❑

29. Which obstacles do you consider to be essential with regard to the development of application field 1 and the exploitation of the market potential? (Multiple references possible)

	unimportant 1	*2*	*3*	*4*	*important* 5
High investment costs	❑	❑	❑	❑	❑
Lack of financing sources	❑	❑	❑	❑	❑
Lack of qualified skilled personnel	❑	❑	❑	❑	❑
Legislation/ Regulation	❑	❑	❑	❑	❑
Lacking market information (Recognition of commercial fields of application)	❑	❑	❑	❑	❑
Lacking technological information	❑	❑	❑	❑	❑
Lacking governmental funding along the whole value-added chain	❑	❑	❑	❑	❑
Lacking availability of competent regional cooperative partners	❑	❑	❑	❑	❑
Below average market potential in Germany	❑	❑	❑	❑	❑
Others ____________________	❑	❑	❑	❑	❑
No serious problems occurring up to now	❑				

Many thanks for your cooperation!

Subject Index

Growth Market Nanotechnology. Edited by N. Malanowski, T. Heimer, W. Luther, and M. Werner

ISBN: 3-527-31457-1